Essener Beiträge zur Mathematikdidaktik

Reihe herausgegeben von
B. Barzel, Essen, Deutschland
A. Büchter, Essen, Deutschland
B. Rott, Köln, Deutschland
F. Schacht, Essen, Deutschland
P. Scherer, Essen, Deutschland

In der Reihe werden ausgewählte exzellente Forschungsarbeiten publiziert, die das breite Spektrum der mathematikdidaktischen Forschung am Hochschulstandort Essen repräsentieren. Dieses umfasst qualitative und quantitative empirische Studien zum Lehren und Lernen von Mathematik vom Elementarbereich über die verschiedenen Schulstufen bis zur Hochschule sowie zur Lehrerbildung. Die publizierten Arbeiten sind Beiträge zur mathematikdidaktischen Grundlagen- und Entwicklungsforschung und zum Teil interdisziplinär angelegt. In der Reihe erscheinen neben Qualifikationsarbeiten auch Publikationen aus weiteren Essener Forschungsprojekten.

Weitere Bände in der Reihe http://www.springer.com/series/13887

Angel Mizzi

The Relationship between Language and Spatial Ability

An Analysis of Spatial Language for Reconstructing the Solving of Spatial Tasks

With a Preface by Prof. Dr. Andreas Büchter

 Springer Spektrum

Angel Mizzi
Essen, Germany

Dissertation der Universität Duisburg-Essen, 2017

Von der Fakultät für Mathematik der Universität Duisburg-Essen genehmigte Dissertation
zur Erlangung des Doktorgrades „Dr. rer. nat."

Datum der mündlichen Prüfung: 27.07.2017
Gutachter: Prof. Dr. Andreas Büchter, Prof. Dr. Benjamin Rott

ISSN 2509-3169   ISSN 2509-3177   (electronic)
Essener Beiträge zur Mathematikdidaktik
ISBN 978-3-658-20631-4   ISBN 978-3-658-20632-1   (eBook)
https://doi.org/10.1007/978-3-658-20632-1

Library of Congress Control Number: 2017963266

Springer Spektrum
© Springer Fachmedien Wiesbaden GmbH 2017

Printed on acid-free paper

This Springer Spektrum imprint is published by Springer Nature
The registered company is Springer Fachmedien Wiesbaden GmbH
The registered company address is: Abraham-Lincoln-Str. 46, 65189 Wiesbaden, Germany

Preface

Spatial-geometrical competencies are required in mathematics continuously, starting from preschool education, then during primary and secondary education until mathematics at university level. Besides arithmetic (which is based on the concept of time), geometry (based on the concept of space) is one of the two initial ways of operating mathematically. The crucial role of geometrical thinking in general and the solving of spatial-geometrical tasks specifically lead to the long research tradition in spatial ability in both mathematics education and psychology.

In traditional psychometric-based intelligence research, spatial-geometrical competencies form a fundamental factor of intelligence in all models. Ordinarily, this factor is considered as largely independent from verbal competencies; especially in popular science discussions, whereby assumptions about different intelligence profiles (e.g., with local strengths in "figural intelligence" or in "verbal intelligence") are sometimes based on such hypotheses.

The interdisciplinary research about the relationship between language and content learning has been researched intensively in Germany in recent years; in mathematics education in particular by Susanne Prediger's research group in TU Dortmund. This research strand emphasises the dependence of other cognitive competencies from the verbal competencies, i.e. the interdependency between verbal and other cognitive competencies. Previous findings in language and content learning research can be categorised either as general results about the integrated verbal and content learning or as results which are specific for particular content in mathematics education. Relevant works are present for mathematical content beyond geometry, e.g. fractions, percentages, descriptive statistics, functions, and elementary algebra. In contrast, there is still a substantial need for research regarding the integrated language and content learning in geometry, in particular in the area of spatial geometry.

A fundamental question arising in the context of mathematics education and mathematics classroom is: Which role do verbal competencies or language use play during the solving of spatial-geometrical tasks? The dissertation of Angel Mizzi addresses this issue - according to the interdisciplinary character of the question - by successfully synthesising theoretical principles from mathematics education, from linguistics and language teaching, and from psychology. The design of the reconstruction method in his work generates a data collection situation which enables a methodologically-controlled empirical study of his underlying research questions. Hereby, a student is required to rebuild a pre-designed object made up of building cubes based on oral instructions from another student. The foci of this study include the strategies developed by students and potential obstacles arising during the solving process, which can be observed in the communication setting. These goals were achieved by an analysis of spatial language used by students in the solving processes.

In view of this theoretical and methodological background, the author succeeds in developing important results which show that language plays an important role during the solving of spatial-geometrical tasks. The analysis of spatial language used by students leads to a differentiation of previous models about strategies used for solving spatial-geometrical tasks. By his own way of use and enrichment of the construct of spatial language, Angel Mizzi succeeds to create an important foundation work for a deeper investigation of the relationship between language and spatial ability. Due to the importance of the underlying issue focussed on in this work further future research projects are expected to emerge following this dissertation.

Prof. Dr. Andreas Büchter

University of Duisburg-Essen

Acknowledgements

I would like to express my sincere gratitude to my supervisor Prof. Dr. Andreas Büchter for his continuous support of my Ph.D study and for his patience and immense knowledge attributing to the intense and fruitful discussions which led to the development of this work. I would also like to thank my second supervisor, Prof. Dr. Benjamin Rott from the University of Cologne, whose guidance helped me at the time of this research and who was always present for answering any questions concerning this research project and mentored my work.

My sincere thanks also goes to the mathematics teachers, especially Daniel Jung, Okan Kaplan and Christina Hohenstein, who helped me conduct the test instruments and select students from their classrooms or schools for this research study.

I would like to thank my colleagues from the Buechter working group at University of Duisburg-Essen, especially Sabine Schlager and Dr. Christina Krause, and overseas colleagues, especially Prof. Boon Liang Chua and Prof. Ban Heng Choy from the National Institute of Education in Singapore and Prof. Wes Maciejewski from San José State University, for supporting me throughout the writing of this thesis.

Last but not least, I would like to thank my family which always supported me throughout my life for achieving a good education.

Table of Contents

Preface V

Acknowledgements VII

1. Introduction..1
   1.1 Overture 1
   1.2 Motivation..1
   1.3 Background and purpose of study..3
   1.4 Aims of research 6

2. Theoretical framework..9
   2.1 Spatial ability 9
      2.1.1 Definitions of spatial ability and approaches to its research.............10
      2.1.2 Cognitive processes in solving spatial tasks 12
      2.1.3 Selected spatial ability models 17
      2.1.4 Spatial abilities in German mathematics curriculum and classroom 22
      2.1.5 Strategies for solving spatial tasks 23
      2.1.6 Sex differences in spatial ability performance 28
      2.1.7 Summary: Spatial ability...30
   2.2 Language in mathematics classroom 31
      2.2.1 Definition of the notion of language 32
      2.2.2 Language and thinking...34
      2.2.3 The functions of language 36
      2.2.4 Levels of language acquisition..37
      2.2.5 Metaphors in mathematics classroom 40
      2.2.6 Roles of language in German mathematics curriculum and
          classroom..42
      2.2.7 Summary: Language in mathematics classroom 44
   2.3 Interplay of spatial ability and language 45
      2.3.1 Spatial language 46
      2.3.2 Spatial task strategies and language 54
      2.3.3 Summary: Interplay of language and spatial ability..........................56
   2.4 Representations of mathematical knowledge...56
      2.4.1 Sfard's dual nature of mathematical conceptions 57
      2.4.2 Bruner's modes of representation 58
      2.4.3 Summary: Representations of mathematical knowledge 60

3. Methodology **61**
   3.1 Research questions 61
   3.2 Research paradigm 63
   3.3 Research design process..66
   3.4 Research method 67
      3.4.1 Reconstruction method...68
      3.4.2 Design principles of the reconstruction method...............70
      3.4.3 Limitations of the reconstruction method 76

4. Design and Implementation **77**
   4.1 Pilot study 77
      4.1.1 Aims of the pilot study 77
      4.1.2 Design of the pilot study 78
      4.1.3 Sampling and implementation...81
      4.1.4 Results and consequences of pilot study 81
   4.2 Design of main study 85
      4.2.1 Reconstruction method in the main study 86
      4.2.2 Task design...87
      4.2.3 Sampling...97
      4.2.4 Quality criteria...108
      4.2.5 Implementation of main study...113
      4.2.6 Data analysis 116

5. Results and discussion from the inductive data analyses....................**125**
   5.1 Description of the identified strategies 125
      5.1.1 Spatial metaphors 125
      5.1.2 Object break-down strategy 142
      5.1.3 Assembling strategy 145
      5.1.4 Rotation strategy 147
      5.1.5 Cubes controlling strategy...149
      5.1.6 Structure controlling strategy 150
      5.1.7 Discussion of the identified strategies and review of literature 151
   5.2 Description of identified obstacles...156
      5.2.1 Spatial metaphors as obstacles 158
      5.2.2 Describing spatial relations 160
      5.2.3 Verbalising rotation..164
      5.2.4 Dimension reduction 169

5.2.5 Space disorientation 171

5.2.6 Discussion of the identified obstacles 172

6. Results and discussion from the deductive data analyses **175**

6.1 Use of identified strategies under consideration of influencing factors..175

    6.1.1 Hypotheses 176

    6.1.2 Language proficiency as a possible influencing factor 178

    6.1.3 Spatial abilities as a possibile influencing factor 181

    6.1.4 Sex as a possible influencing factor 183

    6.1.5 Discussion of the use of identified strategies 184

6.2 Deductive approaches for structural analysis of spatial language and

    influencing factors 187

    6.2.1 Metaphoric-based approach 188

    6.2.2 Linguistic-based approach...195

    6.2.3 Content-based approach 201

    6.2.4 Conception-based approach 209

6.3 Discussion of the results from the structural analyses of spatial

    language...211

7. Summary and Conclusion **215**

7.1 Synthesis of the results...215

7.2 Implications...222

7.3 Concluding remarks 224

References..**227**

1. Introduction

1.1 Overture

The research study reported in this thesis addresses students' use of spoken language for mathematical thinking and learning in the domain of spatial geometry. Two foci of the present work are on the language used to observe and understand students' mathematical thinking and the students' approach for solving spatial tasks. The present study emphasises the importance of spoken language not only to communicate mathematical knowledge, but also to explore and analyse the students' development of mathematical thinking in the domain of spatial geometry, thus establishing a bridge between the domains of language and spatial geometry.

In this chapter, the context and motivation of this study is introduced in Section 1.2. Section 1.3 highlights the importance of the topic in mathematics education and gives a brief overview of results from previous research within this field, which is used to clarify and position the research problem in mathematics education. The aims of this present research are described in Section 1.4 and the last section, Section 1.5, outlines the organisation of the chapters in this thesis.

1.2 Motivation

Before describing the background and stating the aim of the present study in mathematics educational research, I would like to highlight the motivation which triggered this present research work. Firstly, language is not only essential for learning and understanding mathematics at all levels from primary and secondary to tertiary education, but also for the cognitive development of the human being. In the literature, the notion of language learning in mathematics education tends to be associated with the acquisition of new vocabulary. However, throughout this present work the definition of language learning includes the ability to reflect about the language, to be aware of the language depending on the audience, and to be able to successfully adapt to new linguistic situations in a new environment. As a learner of German as a foreign language at the tertiary level during my teaching studies, I have learnt to appreciate the importance of communication to learn and develop one's linguistic and conceptual competen-

1

cies not only for educational purposes, but also to successfully integrate and develop oneself in a new society. Language and content learning is a current issue in the German education system, which is confronted with an increasing number of non-native German speaking learners in mathematics classrooms. The German educational system is currently faced with the need of increasing language awareness and the importance of language in classroom, especially in mathematics and other sciences.

In addition, my collaboration with several teachers in the BiSS project ('Bildung durch Sprache und Schrift' which translates to education through language and writing), a German government funded research programme about developing content and language integrated teaching materials at secondary level, gave me the opportunity to reflect about the use of language in the mathematics classrooms. This collaboration between secondary school teachers and researchers from the University of Duisburg-Essen, Germany, made it possible to get a deeper insight into one of the teachers' challenges to develop language and content teaching materials, which is the understanding and 'grasping' of how students think and how well-developed their language is in order to provide suitable language support in mathematics learning and teaching. In particular, students at the beginning of their secondary education face new challenges arising from the transition from the primary to secondary school. Therefore, from this perspective, this present study gives students a voice so that we, as researchers, teachers and teacher educators, understand more and reflect about student language use, their thinking, and language as a foundation on which both conceptual and language learning can be developed. I think that during the first lessons in secondary schools, when a high amount of new mathematical terms and concepts are introduced, two linguistic environments interact with each other: the teachers' language (which is highly sophisticated) and the students' language. One of the tasks of mathematics teachers is to improve sophistication of students' language. In order to succeed in this, we have to support students to reflect on their language use and create learning moments which show that learning to understand and use mathematical terms and concepts in discourse is worthwhile and necessary for improving mathematics performance. This requires teachers to understand student language and thinking in order to build on their knowledge and identify the improvement potential in language and content learning. The understanding of student's language and their thinking is one of the main aims of this study, which is described in more detail in the subsequent sections.

2

1.3 Background and purpose of study

In this present work, the importance of language in mathematics classroom is demonstrated in the context of spatial geometry. There are several reasons for considering spatial ability and geometry as content for investigating the role of language in mathematics.

Dealing with objects of space and being able to describe them verbally are some of the important skills which students should acquire during their mathematics lessons at primary and lower secondary level (North-Rhine Westphalia Ministry of Education, 2011). Similar to the ability to speak and communicate using one or more languages, the ability to recognise objects in space, the ability to be able to determine one's body position in space and the ability to search for an object in space, are considered to be fundamental experiences to the human species (Landau & Jackendoff, 1993). According to Landau & Jackendoff (1993), one of the main differences between human beings and other species is that human beings are able to express spatial experience and verbalise the structure and location of objects in space. Therefore, from a cognitive point of view, both language and spatial ability seem to belong to the elements of human cognitive structure, which makes their relationship interesting for further research in the field of mathematics education.

Both language (e.g., Clarkson, 1992; Abedi & Lord, 2001; Swain, 2005; Prediger et al., 2013) and spatial ability (e.g., Guay & McDaniel, 1977; Büchter, 2011) are important factors for improving students' mathematics performance. Within the domain of psychology both are considered as separate entities, however, this study shows that these two important domains in mathematics education should not be considered as wholly separate and that they might have more similarities than one might think. Following a systematic search in renomated scientific journals in mathematics education (e.g., Educational Studies in Mathematics (ESM), Journal for Research in Mathematics Education (JRME) and Zentralblatt für Didaktik der Mathematik (ZDM)), no studies have focused on the relationship between language and spatial ability in mathematics education research in the last decades. Therefore, in this work, results from previous mathematics education research in both domains are presented separately, together with the motives for considering the connection of language and spatial ability in research.

3

Language and its role in mathematics education have been gaining more importance in mathematics education research in the last decades (e.g., Pimm, 1987; Moschkovich, 1999; Morgan, 2004; Moschkovich, 2010). In contrast, the importance of language in mathematics classrooms has just been addressed very recently in the German mathematics educational research (e.g., Schütte, 2009; Prediger et al., 2013; Wessel, 2015). This has mostly been triggered by the increasing multiculturalism in German society and the relatively wide gap between German everyday language and German academic language, especially when compared to other Indo-European languages, like English. In particular, the wide gap between everyday and academic language causes a barrier even for German native speaking students in order to achieve high performance in mathematics. These aspects show the need of language-based research in mathematics education to provide not only theories, but to explore the language used in mathematics classrooms to help researchers and teachers understand students' difficulties which might hinder mathematics learning. One of the main aims of this present study is students' use of language to construct mathematical content and reflections about the interplay of language and content, which provides a theoretical basis for language support in future research.

There is a vast amount of literature on language use or multilingualism in mathematics classroom (e.g., Pimm, 1987; Barwell, 2009; Moschkovich, 2010). Whilst several studies (e.g., Maier & Schweiger, 1999; Prediger & Meyer, 2012) have pointed out the functions of language in mathematics education, few if any researchers have addressed the interplay of communication and cognitive processes in the context of spatial geometry. Communication is vital for the learning and teaching of mathematics, however, the nature of the cognitive processes which are influenced by language to construct spatial knowledge have not been dealt with in depth in previous mathematics education research.

In the field of mathematics education, much of previous research on spatial ability (e.g., Smith, 1964; Fennema, 1974; Guay & McDaniel, 1977; Büchter, 2011) has strongly focussed on the assessment of students' spatial abilities mostly by using written tasks, which has been influenced by earlier psychology research about spatial ability. A key problem with much of the literature about spatial ability is the negligence of considering communication as important tool to help understand spatial thinking of students whilst solving spatial tasks. Whereas the use of pencil-and-paper tests for assessment of students' spatial abilities is widely used in previous studies (e.g., Büchter, 2011), very little is known about the use of spoken language in solving spatial tasks in mathematics education. In her recent study about spatial ability, Plath (2014) analysed student language during solving spatial tasks by requiring the students to verbalise their thinking to explain their reasoning and argumentation in spatial tasks. A wide collection of students' strategies is described in Plath (2014), as well as the relationship between spatial ability and mathematics performance. However, the role of language in her research is not strongly emphasised in her approach, especially the nature of students' language used in solving spatial tasks.

Geometrical thinking and knowledge is vital for models of spatial ability (e.g., Pinkernell, 2003) and this will be considered as crucial in the underlying definition of spatial ability in this present study. Whilst I agree that written tasks, such as paper-and-pencil tests are important to assess students' spatial abilities, they are not sufficient to understand how students solve spatial tasks and which processes are dominant during the solving of such tasks. Therefore, the aim of this present work is to design a study which integrates language in the process of solving spatial tasks for understanding better how students solve these tasks and which obstacles they encounter during the task solving process. This present study shows that previously described strategies in solving spatial tasks, such as *analytic* vs. *holistic* (cf. Barrat, 1953), *three-dimensional* vs. *two-dimensional* (cf. Gittler, 1984) or *rotators* vs. *non-rotators* (cf. Geiser, Lehmann & Eid, 2006) are not sufficient to describe strategies in solving specific spatial tasks requiring the use of language.

When analysing the strategies used during solving of spatial tasks, one has to consider other factors which might influence the use of strategies and performance in solving spatial tasks. When considering the students' use of language during solving spatial-geometrical tasks, students' language proficiency and spatial abilities are the first two plausible factors which might influence the solving of spatial-verbal tasks. Prior research (e.g., Linn & Petersen, 1985; Büchter, 2011) about spatial ability points out that sex is another important factor for solving spatial tasks, which should not be underestimated. Therefore, additionally, this study examines to which extent these three factors – language proficiency, spatial abilities and sex – could play a major role in students' use of strategies when solving spatial geometrical tasks.

The interaction between language and spatial ability should not only be investigated by focussing on strategies and obstacles when students describe spatial objects. Another important aspect is the nature of students' spatial language in solving spatial tasks, which in most literature is referred to as the language used to describe spatial objects. Spatial language is a neglected area in the field of mathematics education. However, it seems to be a promising research area to support the importance of investigating the relationship between spatial ability and language. Therefore, another goal of this present study is to explore typical characteristics of students' spatial language used in solving spatial tasks in order to broaden our knowledge about this type of language and help us understand student's spatial thinking from a mathematics educational perspective.

1.4 Aims of research

The main aim of this present study is to investigate how students solve spatial geometrical tasks by focusing on the use of language in the task solving process. Therefore, the two interrelated main research questions are:

1) **How do different students solve spatial-verbal tasks?**
2) **What role does language play in the description of spatial configurations in spatial tasks?**

To provide empirical answers for these global research questions, different aspects of solving spatial tasks and spatial language need to be considered. The first goal of this thesis will be to analyse and classify the different strategies used by the students to solve spatial-verbal tasks. Second, possible obstacles emerging during the task solving process will be identified and described in detail. Third, the use of strategies in the solving process will be analysed regarding any dependency from language proficiency, spatial abilties, and sex of the students. In the next research aim, spatial language will be analysed in order to provide a deeper insight in the linguistic elements carrying spatial meaning and their use among students under consideration of the above-mentioned factors. From a methodological point of view, another goal of this present study is the theoretical and empirical description of a suitable method for investigating the use of language to solve spatial tasks. The research questions corresponding to these five goals is presented at the beginning of the methodology chapter, Chapter 3, following an introduction and a thereotical discussion of the concepts and notions in Chapter 2.

1.5 Organisation of the thesis

An overview of the theoretical background of this thesis is provided in Chapter 2. The theoretical background section addresses the theoretical basis of this present study, which includes results from previous research about language in mathematics education, spatial ability and spatial language. In the first part, the concept of spatial ability and its different models in psychology and mathematics education, its underlying cognitive processes, different strategies used in solving spatial tasks, and sex differences in spatial ability performance are discussed. In the second part of the theoretical background, the concept of language is introduced, together with functions of language and varieties of language in mathematics classrooms. The next part of the chapter is dedicated to spatial language, which can be depicted as the interaction between language and spatial ability. Further theoretical background about other concepts which are relevant to the topic of this present work, such as the different representations of mathematical (spatial) knowledge, is also presented.

In the methodology chapter, Chapter 3, the research paradigm and the research design process of this present study is illustrated and discussed. In the first part of Chapter 3, the research questions corresponding to the aims of this present study is introduced. This is followed by the description of the research paradigm and the research design process. The domains of language and spatial ability will be interlaced in a method of data collection, called 'reconstruction method', which until now has only been applied in a few studies if any. The reconstruction method will be described and justified as an adequate qualitative research method on a theorical level.

In Chapter 4, the design and implementation of this present study is discussed. The first part of this chapter is dedicated to the pilot study, which played an important role in designing the main study. This is followed by a detailed description of the main study, including the design of tasks, the description of the sampling group and instruments for sampling, the implementation of the study, data collection and analysis.

Chapter 5 presents the results concerning the qualitative research aims of this present study and their discussion, i.e. the description of strategies used by students and the obstacles encountered during the solving of the spatial-verbal tasks. In the first part of Chapter 5, the different strategies observed during the data analysis are described and justified by referring to previous literature. The second part of this chapter provides results concerning students' obstacles in describing spatial objects, in which different case studies are presented in order to describe the different students' obstacles identified during the solving of the spatial-verbal tasks.

Chapter 6 is dedicated to explorative analyses of qualitatively collected data of students solving spatial-verbal tasks for investigating the frequency of use of identified strategies in students' solving processes and systemic analyses of student spatial language. In the first part of Chapter 6, the identified strategies described in the first section of Chapter 5 are analysed to observe how they vary with other background factors, such as students' language proficiency, spatial abilities, and sex. The second part of this chapter provides an exploration of student spatial language by providing different approaches for analysing students' spatial language under consideration of the above-mentioned background factors. Results concerning patterns observed in student spatial language used during the spatial task solving processes are presented and discussed in Chapter 6.

Finally, Chapter 7 provides a summary and synthesis of the findings of the present study. This is followed by implications for future research and for teaching in the field of language and content learning in mathematics education.

2. Theoretical framework

In analysing how students[1] solve spatial geometrical tasks, I am interested in how language can play a role in the students' solving process. In a wider context, I intend to study the relationship between language and spatial ability for a deeper understanding of language and content learning in mathematics education. This section provides an overview of the theoretical frameworks in both domains of language and spatial ability, followed by theoretical considerations on the integration of both domains based on findings from previous literature.

In Section 2.1, the theoretical findings about spatial ability from psychology and mathematics education are presented. In order to better understand the concept of spatial ability and to account for the design of the main study later in this work, selected models of spatial ability and other important results from previous studies about spatial ability are discussed. The major findings about language and its use in mathematics learning and teaching from previous research, which includes the functions of language, the use of metaphors and different levels of language acquisition in mathematics classroom are presented in Section 2.2. In Section 2.3, the notion of spatial language as an interface between language and spatial ability is introduced by referring to important results from psycholinguistic studies. Theoretical approaches to the representation of mathematics knowledge, such as Bruner's stages of representation and Sfard's theoretical framework for investigating the role of algorithms in mathematical thinking, are presented in Section 2.4.

2.1 Spatial ability

Traditionally, spatial abilty has been extensively researched in the field of psychology (e.g., Thurstone 1950; Linn & Petersen, 1985). However, the past two decades have seen a renewed importance of spatial ability in mathematics education (e.g., Pinkernell, 2003; Casey et al., 2008; Büchter, 2011; Plath, 2014). The considerable amount of literature on spatial ability led to the establishment of a wide spectrum of definitions of spatial ability, some of which are introduced in the first part of this section. The next section, Section 2.1.2, addresses the cognitive processes playing an important role in solving spatial tasks. Based on the

---

[1]   In the rest of this thesis, the terms 'students' is used to refer to students at primary or secondary level.

different definitions of spatial ability, different models of spatial ability from psychology and mathematics education are introduced in Section 2.1.3. In Section 2.1.4, the role of spatial ability in mathematics classroom and its relationship to geometry learning and teaching based on theoretical discussions from previous literature and based on German mathematics curriculum are discussed. This is followed by the discussion of different strategies which, according to previous literature, students use during solving of spatial tasks in Section 2.1.5. The last part of this section, Section 2.1.6, is dedicated to the influence of sex on solving spatial tasks based on previous reported empirical results.

2.1.1 Definitions of spatial ability and approaches to its research

One can find several definitions of the term *spatial ability*[2] in the literature, but spatial ability seems to be a collection of abilities for the manipulation of space, which are required to solve spatial tasks. Similar to the notion of problem in problem solving (cf. Pólya, 1949/1980; Kantowski, 1980; Carpenter, 1988), in this work, spatial tasks will be considered as non-routine tasks demanding spatial ability and other abilities and as tasks which require students to develop ways to overcome the obstacles in space in order to reach the goal of the spatial task. Due to the interchangeable use of the notions of *spatial ability, spatial knowledge, spatial intelligence, spatial thinking* and *spatial reasoning* in previous literature, a brief distinction between these notions will be undertaken for a better clarification of their use in this work. Whereas spatial ability encompasses different abilities required to navigate in space and solve spatial tasks (both in mental and real terms), spatial knowledge (perhaps also spatial intelligence often used in psychometric approach to spatial ability) seems to emphasise primarily the spatial information and positions in space stored and activated in the individual's cognitive system, whereby it can be stored using multiple representations (e.g., pictorial or symbolic). However, spatial knowledge can also encompass knowledge about strategies of dealing with spatial tasks and navigating in space from previous experience in solving spatial tasks. Whereas spatial reasoning emphasises the use of logic for solving spatial tasks, spatial thinking encompasses mental processes characterising the solving of a spatial task, in which students apply spatial reasoning and are required to develop strategies for reaching the goal addressed by the underlying spatial task.

---

[2] The plural of spatial ability, *spatial abilities*, emphasises the different components in spatial ability models.

Pinkernell (2003) defines spatial ability as the ability to mentally operate on spatial objects in space. Similarly, Thurstone (1938) views spatial ability as the ability to mentally operate on two and three-dimensional objects. In their definition of spatial ability, Linn and Petersen (1985) define spatial ability as referring to "skill[s] in representing, transforming, generating, and recalling symbolic, nonlinguistic information" (Linn & Petersen, 1985, p. 1482). Maier (1999) considers spatial ability not merely as the cognition and perception of visual information, but also mental processing of this information should belong to the domain of spatial ability. He defines spatial ability as the ability to view mental images in space and to think spatially (cf. Maier, 1999). The development of mental images, which do not necessarily require a reference object in real life, is an important spatial ability, however, a person with high spatial ability should also be capable to use these mental images and rearrange these mentally for developing new mental images from the ones available (cf. Maier, 1999).

From a mathematics education perspective, the notion of spatial ability should encompass not only the mental representation of space and its transformation in terms of Maier (1999) or Pinkernell (2003), but also in real actions. This includes spatial tasks which might also require motoric skills, such as navigating through a labyrinth or streets of a city, or moving or rotating a spatial object, or folding paper to solve particular spatial tasks. Hence in this present study, spatial ability shall be defined as a collection of abilities which change or support our perception of space and can be represented either mentally or in real terms using different representations of spatial knowledge.[3] The term *spatial abilities* will be used to emphasise the different abilities in spatial ability, especially concerning the assessment of spatial ability (see Section 2.1.3).

Similar to the different definitions of spatial ability, there are different approaches to and perspectives in research about spatial ability, depending on the underlying aims and goals of research. Linn and Petersen (1985) differentiate between four different research perspectives of spatial ability – strategic, cognitive, psychometric and the differential perspective. In the strategic research perspective, the test persons' strategies during the spatial task solving process are investigated qualitatively and identified, whereas the goal in the cognitive perspective is to identify and characterise cognitive processes during the spatial task solving pro-

---

[3]   In contrast to Linn & Petersen's (1985) definition of spatial ability, in this thesis, spatial abilities can also consist of recalling linguistic or symbolic information.

11

cess (cf. Linn & Petersen, 1985). In the cognitive perspective, mental processes during the spatial task solving process are more important to characterise spatial abilities than the test results (cf. Linn & Petersen, 1985). In the third research perspective, the psychometric perspective, spatial ability is divided into subfactors or components by comparing correlations between different spatial tasks (cf. Linn & Petersen, 1985). Psychometric research was prominent among psychologists (e.g., Thurstone, 1938) who describe different factors of spatial ability in their models (see Section 2.1.3.3). In the differential perspective, research is characterised by a "comparison of spatial ability for different populations" (Linn & Petersen, 1985, p. 1480), whereby the performance in spatial ability tests are indicators for differences in spatial ability among different groups, for example, males versus females (cf. Linn & Petersen, 1985).

The foci in this present study include strategies and obstascles of students during the solving of spatial-verbal tasks which will be identified by analysing the use of language in order to understand how students solve spatial tasks. This means that the strategic and cognitive perspective of research on spatial ability are the most relevant for this present study.

2.1.2 Cognitive processes in solving spatial tasks

There are several cognitive processes which are activated when students solve spatial tasks. In her study about how fourth-grade students solve spatial tasks, Plath (2014) states that perception of visual information, processing of information, generation of mental images and thinking are the most important cognitive processes taking place during the spatial task solving process. These cognitive processes should not be analysed as separate, but rather as mutually influential and as building up on each other, preferably in a cycle in the order they are introduced in the following sub-sections, to construct a complex system of cognitive processes (cf. Plath, 2014).

2.1.2.1 Visual perception

In general, the process of perception can be described as a process in which the person becomes aware of a particular knowledge in a particular environment. According to Lurija (1992), in modern psychology, perception is an active process consisting of several steps to be mastered:

It [modern psychology] describes perception as an active process, in which information is searched, unique characteristics of an object are identified, these characteristics are compared to each other, adequate hypotheses are formulated and comparisons among these hypotheses and the initial data are made. (Lurija, 1992, p. 230)[4]

According to Frostig, Horne and Miller (1977) there are different types of visual perception. The most perceptive skills relevant for this present study are figure-ground perception, spatial relations perception and spatial position perception, which are explained below (the other ones can be found in Maier, 1999).

a) *Figure-ground perception*: This perceptive process deals with the ability to identify a figure or an object from an optical complex background or to identify parts of the object from a given object and isolate them (cf. Maier, 1999).

b) *Spatial relations perception*: The visual perception of spatial relations is the ability to visually identify and describe the relation between two or more spatial objects, for example, the ability to describe the spatial relation between two solids in geometry (cf. Maier, 1999).

c) *Spatial position perception*: In the perception of position in space, an individual shows his or her ability to identify and perceive the position of an object in space under consideration of his or her own body. Maier (1999) describes this perception as being close to the perception of spatial relations, the only difference being the importance of the body position for solving the spatial task successfully. The main difference between spatial position and spatial relations is that in spatial position one of the objects to which it is related to must be the observer's body.

Hence, visual perception in space is the perception of spatial information from the surrounding environment, which requires selection of information of specific spatial or geometrical characteristics of spatial objects, or spatial relations between two objects, or the spatial position of one's own body in space and relate it other spatial objects. The perception of such spatial information is needed for the creation of mental images in the cognitive system, which is discussed in the next section.

---

[4]   This quotation was translated from German to English and is reproduced without any further alterations from Lurija (1992, p. 230) by A.M.

2.1.2.2 Mental images

The notion of *mental images* and their formation is an important cognitive process when solving spatial tasks, and is the subject of different definitions in literature. Mental images are described as "the mental invention or recreation of an experience that in at least some respects resembles the experience of actually perceiving an object or an event, either in conjunction with, or in the absence of, direct sensory stimulation" (Finke, 1989, p. 2). Finke (1989) states that mental images are spatially arranged as objects are arranged in space, and that "imagined transformations and physical transformations exhibit corresponding dynamic characteristics" (Finke, 1989, p. 93). However, not every mental image, as an abstraction of a perceived external stimulus, does necessarily require a real object which corresponds one to one to it; as in the example of a mental image of a blue horse, which is formed by combining the images of 'blue' and 'horse' together to create a new mental image. Therefore, the process of mental image creation is an active process which enables the creation of new mental images from existing knowledge and enables new produced mental images to be used for the individuals' thinking and acting in the environment (cf. Plath, 2014).

2.1.2.3 Information processing

The processing of information during solving of spatial tasks is another important cognitive process following visual perception. There are two different approaches in psychology to explain how information is processed in the mind: propositional-code (cf. Pylyshyn, 1973) and dual-code theories (cf. Paivio, 1971). In the former theory, Pylyshyn (1973) proposes that information is coded as prepositions, which are defined as the smallest units of knowledge, and contain a relationship called the predicate and a number of arguments. Pylyshyn (1973) argues that propositions should not be equated with a string of words, and their representation in the mind cannot be solely described by pictures or words:

> A proposition is what a string of words may assert. A proposition is either true or false; the string of words is neither. A proposition may be asserted by a number of strings of words, in any language and in any modality. Furthermore, in the sense in which we use it when we speak of "propositional knowledge," it may involve no words of any kind. Thus when I look at the table in front of me, I see that there is a vase on it. I do not "see" patches of light or only an array or objects. My knowledge is en-

riched by (among other things) the proposition asserted by a sentence such as "The vase is on the table", even though I did not utter (audibly or otherwise) this or any other sentence. (...) When we use the word "see," we refer to a bridge between a pattern of sensory stimulation and knowledge which is propositional. (Pylyshyn, 1971, p. 6)

Hence, Pylyshyn's (1971) model supports the idea that information is processed in a system of propositions, irrespective of whether the information is visual or verbal. In contrast, Paivio (1971) introduces a more widely acknowledged model of information processing which differentiates between the two kinds of information processing for verbal and imagery, the dual-code theory. The dual-code theory states that nonverbal visual information is processed in the image system and verbal information is processed in the verbal system (cf. Paivio, 1971). However, whilst both systems are not dependent on each other, they are interconnected (cf. Paivio, 1971). The following example should clarify the relationship between both systems in the dual-code theory. Consider an individual who is watching an elephant on television, this does not necessarily mean that he or she produces a mental image in the image system, but it could activate verbal information in the verbal system. Reciprocally, if an individual is exposed to the word 'elephant' it might lead to the activation of more verbal information and perhaps to the production of a corresponding mental image (cf. Plath, 2014). However, if the verbal information is very abstract, the individual is less likely to produce a mental image of the corresponding verbal information (cf. Paivio, 1971). According to Paivio (1971), information processes consist not only of the production of mental images or verbal information, but also of the connection between the image and verbal system. In terms of spatial information processing, Pavio's (1971) dual theory captures two important planes for representing and hence most likely also for processing spatial information: the visual level, which is essential for space perception and for storing mental images of spatial constellations or transformations which are not yet verbally processed, and the verbal level, which is vital for abstaction of spatial knowledge and its verbalisation.

2.1.2.4 Thinking process

After the perception of spatial objects on a visual level, processed information arising from mental images requires thinking as a cluster of mental actions for real actions (cf. Aebli, 1980). In her neuropsychological approach to the notion of thinking, Lurija (1992) describes thinking as a set of processes which can be categorised in three phases. According to Lurija (1992), thinking processes are triggered by the subjects' need to consider a task or problem and find a solution for it urgently, and when the subject is confronted with a situation, which does not have any ready solution (innate or habitual). In the second phase of thinking processes, subjects analyse the task requirements and components, identify the most important characteristics and compare these with each other (cf. Lurija, 1992). The third phase, also known as strategic thinking, consists of the subject's choice of a solution among different possibilities, and the development of a plan for finding the solution of the problem (cf. Lurija, 1992). Hence, the phases of thinking processes in Lurija (1992) coincide with the notion of problem solving (i.e. development of strategies) and metacognition in mathematics education.

Based on Freud's (1982) definition of thinking as inner experimental test action, Spering and Schmidt (2009) point out that thinking, which is defined as mental monitoring of reality, does not necessarily induce real actions, since an action developed from thinking processes can be performed mentally. The objects of thinking processes or thoughts are dependent on the individual and his or her knowledge, and on the context or task in which the subject needs to make up a plan, decide, and reason to solve a task (cf. Spering & Schmidt, 2009). However, objects of thoughts can also be thoughts themselves, whereby the subject is able to analyse his or her way of thinking and analyse thinking processes to process more compex reasoning[5] or search for errors in reasoning (cf. Spering & Schmidt, 2009).

In this section, four cognitive processes which are essential for solving spatial tasks have been introduced. However, this should not exclude that other processes can play an important role in solving spatial tasks. For instance, in developmental psychology, language is regarded to be a major factor influencing thinking processes. Vygotsky (1993) considers language as an inner speech which is regarded as a medium for supporting thinking processes. A detailed account of the relationship between thinking processes and language is discussed in Section 2.2.2. Vygostky's (1993) approach to thinking process supports the assumption that thinking processes should not be analysed separately from the other cognitive processes. However, in a spatial context, thinking processes should be an essential aspect for processing visual or verbal spatial information and manipulat-

---

[5]   In this thesis, thinking denotes all thinking process which are produced either conscious or unconscious, whereas reasoning should refer to conscious thinking processes which involve the use of logic.

ing it, whether the leaner uses mental or real actions to reach the goal stipulated by the underlying spatial task. Another factor that influences spatial thinking is the subject's spatial knowledge or several spatial abilties, which are discussed in the upcoming section.

2.1.3 Selected spatial ability models

As mentioned in the first section of this chapter, spatial ability has been an object of research among both psychologists and mathematics educators. Due to the earlier extensive research in psychology (mainly starting during the 1930s), some models of spatial ability in mathematics education are strongly influenced by the psychological approach. For a better understanding of the diversity in spatial abilities, four models of spatial ability, two from the psychological perspective (cf. Thurstone, 1950; Linn & Petersen, 1985) and two from the perspective of mathematics educators (cf. Maier, 1999; Pinkernell, 2003), chosen due to their clarification and adequacy of the notion of spatial abilities in previous literature (e.g., Büchter, 2011), are introduced in the upcoming subsections. As already mentioned in Section 2.2.1, spatial ability is considered as a component of intelligence in psychology, which is dominated by the need to describe and classify its inner structure and its complexity into sub-factors or spatial abilities (e.g., Thurstone, 1950). In contrast, recent spatial ability models (e.g., Pinkernell, 2003) in mathematics education tend to focus more on the role of spatial ability in learning and teaching mathematics, especially in geometry classes. This comparison shows the wide spectrum of definitions and competencies associated with the notion of spatial ability in psychology and mathematics education.

2.1.3.1 Thurstone's spatial ability model

Thurstone's (1938) work on intelligence theory, in which he formulated the different primary components of intelligence in *principal landmarks in mental ability* (cf. Thurstone, 1938) was an important initial step to emphasise and classify the psychological core of spatial ability. In his factor model, space or spatial abilities are considered as one of the primary mental abilities, together with other factors such as verbal, word fluency, number, perception, memory and reasoning (cf. Thurstone, 1938).

In his work about investigating test persons in terms of abilities in visual thinking, Thurstone (1950) was also one of the psychologists to offer a model of spatial ability based on the differentiation between three factors of spatial abilities:

visualization, spatial relations and *spatial orientation*. *Visualization* is defined as the "ability to imagine the movement or internal displacement among the parts of a configuration that one is thinking about" (Thurstone, 1950, p. 518). The factor *spatial relations* is "the ability to recognize the identity of an object when it is seen from different angles" (Thurstone, 1950, p. 518). Thurstone (1950) states that wheras the objects or configurations are rigid in the factor of spatial relations, the factor of visualization implies movement of object or configuration. Finally, Thurstone (1950) defines *spatial orientation* as the ability to "think about those spatial relations in which the body orientation of the observer is an essential part of the problem" (Thurstone, 1950, p. 519). Thus, in contrast to spatial relations, in spatial orientation the position of the person solving the spatial task is important for the successful accomplishment of the underlying task.

Thurstone's (1950) spatial ability model was one among the first to describe the different factors of spatial ability and integrate them in a model. It served as a basis for extensive discussion about the nature of spatial abilities and influenced other spatial ability models, not only in psychology, but also in mathematics education (e.g., Maier's model of spatial ability in Section 2.1.3.3).

2.1.3.2 Linn and Petersen's spatial ability model

Linn & Petersen (1985) offer another approach to the structure of spatial abilities by focusing on the different strategies students employ during the solving of spatial tasks. In their meta analysis consisting of 172 studies, they presented three components of spatial abilities and reference tests, which are used to measure the corresponding spatial ability (see Table 1). As Table 1 shows, this model is based on the three components – *spatial perception*, *mental rotation* and *spatial visualization* – which denote orientation by using the gravitational vertical or horizontal, mental rotation and multistep analytic procedure respectively.

One of the advantages of Linn & Petersen's (1985) model are the straightforward definitions of each component and the measurement of each component via the corresponding cognitive tests (see Table 1). Büchter (2011) describes Linn and Petersen's (1985) model as cognitive-psychologically and psychometrically plausible and uses this model together with the corresponding reference tests to measure students' spatial abilities to investigate the relationship between spatial ability and mathematics performance. This model is used not only to describe the concept of spatial ability, but also to provide assessment of student spatial abilities through reference tests.

Component	Definition	Reference test
Spatial Perception	Spatial perception is defined as the spatial ability "to determine spatial relationships with respect to the orientation of their own bodies, in spite of distracting information." (Linn & Petersen, 1985, p. 1482)	Water Level Task (WLT) (cf. Piaget & Inhelder, 1958); Rod and Frame Test (RFT) (cf. Witkin, Dyk, Faterson, Goodenough & Karp, 1962)
Mental Rotation	Mental rotation is the spatial ability to "rotate a two or three dimensional figure rapidly and accurately" (Linn & Petersen, 1985, p. 1483)	Mental Rotation Test (MRT) (cf. Linn & Petersen, 1985); Flags and Cards (c.f. French, Elstrom & Price, 1963)
Spatial Visualization	Spatial visualization is the component "associated with those spatial ability tasks that involve complicated, multistep manipulations of spatially presented information" (Linn & Petersen, 1985, p. 1484).	Different Aptitude Set – Spatial Relations Subtest (DAT:SR) (cf. Bennett, Seashore & Wesman, 1973); Paper Folding (cf. Linn & Petersen, 1985)

Table 1   Components of Linn and Petersen's (1985) spatial ability model

2.1.3.3 Maier's spatial ability model

Psychological spatial ability models have also served as a basis for mathematics educators researching the domain of spatial ability. Maier's (1999) spatial abilty model integrates spatial-visual components of different psychological models (e.g., Thurstone, 1950; Linn & Petersen, 1985). Maier's (1999) model consists of six spatial-visual components: *spatial perception* and *mental rotation* according to Linn and Petersen (1985) (see Section 2.1.3.2) and *spatial visualization, spatial relations*, und *spatial orientation* according to Thurstone (1950) (see Section 2.1.3.1) and *factor K*. The last component, *factor K*, is based on the factor kinesthetic imagery in Michael, Guilford, Fruckter and Zimmermann (1957)'s three factor model of spatial ability. They define it as representing "merely a left-right discrimination with respect to the location of the human body" (Michael et al. 1957, p. 191 as citied in Maier, 1999, p. 45).

Regarding the importance of spatial abilities in mathematics classroom, Maier (1999) emphasised that the three components, *visualization, spatial relations* and to greater extent *spatial orientation* are the most relevant in mathematics classroom (as well as in everyday life and work). The importance of these three factors is also visible in his model, in which the areas representing these factors are enlarged (see Figure 1). The smaller areas in the model representing the other factors of spatial abilities (*mental rotation, spatial perception* and *factor K*) in the model have been described as rather special and specific (cf. Maier, 1999).

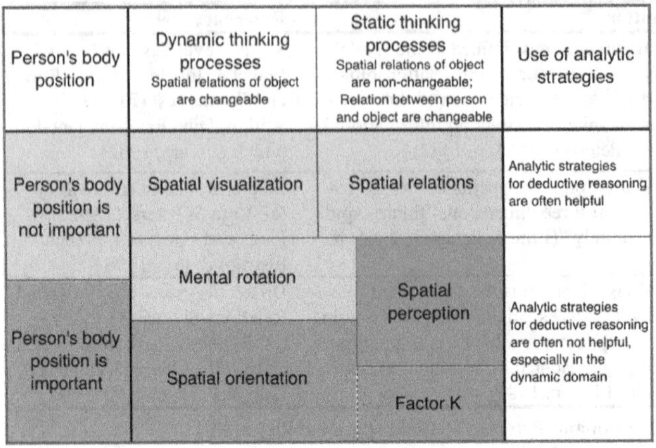

Person's body position	Dynamic thinking processes Spatial relations of object are changeable	Static thinking processes Spatial relations of object are non-changeable; Relation between person and object are changeable	Use of analytic strategies
Person's body position is not important	Spatial visualization	Spatial relations	Analytic strategies for deductive reasoning are often helpful
Person's body position is important	Mental rotation Spatial orientation	Spatial perception Factor K	Analytic strategies for deductive reasoning are often not helpful, especially in the dynamic domain

Figure 1 A visualisation of the spatial ability model called "the (spatial) map" according to Maier (1999, p. 71)[6]

Maier's (1999) model shows the acknowledgment of psychological spatial ability models in mathematics education, especially how psychological approaches to spatial abilities can be integrated in mathematics education. In particular, Maier (1999) shows the importance of Linn and Petersen's (1985) model in mathematics education and considers Thurstone's (1950) notion of spatial orientation, which have been neglected or not explicity mentioned in other spatial ability models in mathematics education. Based on this model, Maier (1999) also shows how spatial abilities are important in geometry by describing which spatial abilities are required for which geometrical task (see Section 2.1.4). However, due to the lack of consideration of geometrical knowledge explicitly in his model, another spatial ability model in mathematics education is introduced in the next section.

2.1.3.4 Pinkernell's spatial ability model

Pinkernell (2003) developed an integrative model for spatial abilities after concluding that the psychological models are not enough to deliver appropriate results in the field of mathematics education. Pinkernell's model describes various spatial abilities which are allocated in three categories, *spatial-visual operations* (*Räumlich-visuelles Operieren*), *geometrical thinking* (*Geometrisches Denken*)

---

[6]   The diagram in Figure 1 was translated from German to English and is reproduced without any further alterations from Maier (1999, p. 71) by A.M.

and *visual abilities (Visualisierungskompetenz)*. *Spatial-visual operations* consist of mental actions on spatial objects and real actions in space. Abilities belonging to spatial-visual operations include the ability to mentally construct or reproduce objects, to transform their spatial-visual characteristics, for example, their position or form, to model, to draw and to describe real space object and its transformation (cf. Pinkernell, 2003). Pinkernell's (2003) second category of spatial abilities, *geometrical thinking*, includes abilities such as recognising and understanding spatial objects, describing them and their transformation under consideration of their geometrical characteristics. Abilities within this category include the reduction of solids to their geometrical form, pattern or structure to facilitate problem-solving processes. The last category, *visual abilities*, is the interpretation and construction of different representation forms of spatial-visual objects, such as models, graphs and the description of configurations using the correct terms for the underlying concepts (cf. Pinkernell, 2003).

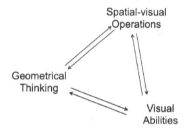

Figure 2   A visualisation of Pinkernell's (2003, p. 57) spatial ability model[7]

Therefore, Pinkernell's (2003) model provides a collection of abilities which can be directly associated with in the solving of spatial task in mathematics education. The first category in this model addresses both internal and external representations of spatial objects; the latter having been ignored in psychological models of spatial ability. Additionally, the second category deals with geometrical thinking, which plays a major role for the application and fostering of spatial abilities in mathematics classrooms. The different representation forms and communication, which are essential for teaching and learning mathematics, are also addressed in the last category of Pinkernell's (2003) model. This model rep-

---

[7]   The diagram in Figure 2 was translated from German to English and is reproduced without any further alterations from Pinkernell (2003, p. 57)

21

resents the notion of spatial abilities in this thesis due to Pinkernell's (2003) success in integrating both fundamental elements from psychological and mathematics education models about spatial abilities. It is suitable from the perspective of mathematics education since it allows an analysis of student behavior in solving of spatial tasks and highlights the importance of spatial abilities in the interplay among the three categories (see Figure 2).

2.1.4 Spatial abilities in German mathematics curriculum and classroom

The learning and training of spatial abilities is an important part of geometry lessons in mathematics. However, their development and training in mathematics classroom is not directly addressed in mathematics curriculum, but rather represented in conjunction to knowledge, competencies, or tasks in spatial geometry. Examples of such competencies formulated in the German mathematics curriculum include students' ability to produce nets of solids, to mentally operate with solids, and to classify geometrical objects (cf. Federal German Ministry of Education, 2004). The reluctance to address spatial abilities explicitly in mathematics curriculum leads to the assumption that spatial abilities seem to be more of a requirement for geometry learning in mathematics lessons, rather than a topic to learn by itself.

Maier (1999) provides an extensive summary of geometrical content which shows the development of spatial abilities in spatial and plane geometry in mathematics classrooms according to his model (see Section 2.1.3.3). Spatial-geometrical tasks which belong to the factor of spatial perception include discovering of symmetry of solids or figures and identifying solids' or figures' height (cf. Maier, 1999). Spatial abilities within the component of visualization and spatial relations can be developed by using building cubes to build solids, describing everyday-life objects and recognising basic shapes, drawing or folding geometrical figures, solids and eventually concepts (e.g., parallel lines), estimating and comparing length, areas and volume etc. (cf. Maier, 1999). Adequate geometry topics for Maier's (1999) spatial componenent of rotations are dealt with at secondary level, for instance, when introducing the concepts of rotation, inversion in a point and solids of revolution. The fourth spatial component of spatial orientation can be developed in geometry by expressing spatial relations between locations of objects and persons (e.g., to the right or left of...) at primary level, and orientation in a grid coordinate or in trigonometry at secondary level (cf. Maier, 1999).

The above-mentioned tasks in geometry lessons show how spatial abilities and their development are integrated within geometrical learning in mathematics classes in primary and secondary schools. The integration of spatial abilities in geometry classroom has been an issue in mathematics education for a long time. For instance, Ilgner (1974) argued that the development of spatial abilities in mathematics classrooms is not only the focus in particular geometry topics or tasks, but it should be rather considered as a methodological principle for geometry lessons in general to support both geometrical and spatial thinking in mathematics. The consideration of geometrical thinking as part of spatial abilities opens new perspectives to the approach to spatial abilities, which have been neglected in spatial ability models adopted from psychology (e.g., Linn & Petersen, 1985). As Pinkernell's (2003) model shows, abilities such as constructing models and describing properties or symmetries of spatial objects are also important for the research, development and assessment of student's spatial abilities in mathematics education. As a matter of fact, Pinkernell's (2003) model will be considered as an important theoretical framework for the notion of spatial abilities in this present research.

2.1.5 Strategies for solving spatial tasks

When solving spatial tasks, students need to develop or choose among strategies to solve the tasks, which demand particular spatial abilities.[8] The existence of different strategies for solving spatial tasks has also been recorded in previous studies about strategy choice in solving spatial tasks: "the preceding brief review of the processing strategies used by individuals on standard spatial tests makes it clear that different individuals attempting the same spatial task do, in fact, use a wide range of strategies" (Clements, 1983, p. 15).

After a brief introduction of the notion of strategy in the first part of this section, different strategies for solving spatial tasks from previous research is presented. The selected strategies for solving spatial tasks include Barrat's (1953) analytic and holistic dichotomy and Gittler's (1984) three-dimensional and two-dimensional thinking.

---

[8]   The fact that solving of spatial tasks requires strategies rather than just activation of knowledge indicates the similarity between the notion of spatial task in the present study to the notion of (spatial) problem in the domain of problem-solving.

2.1.5.1 Definition of strategies

The term *strategy* is often used in the domain of problem solving in mathematics education, but few mathematics educators give an extensive definition of this concept. Schoenfeld (1983) describes two types of decision moments that can take place during the process of problem solving, which can be considered as building the core of his notion of strategies. These moments are induced by two main questions: *what shall be done to solve the problem?* and *how can this be achieved to solve the problem?* (cf. Schoenfeld, 1983 as cited in Fülöp, 2015). In the former ones, which are called strategic decisions, aims and goals are formulated and decisions are taken via actions. In the latter types of decisions, the focus is on the implementation of the decisions which have been taken in strategic decisions (cf. Fülöp, 2015). Based on Schoenfeld (1983)'s previous work, Fülöp (2015) provides a more explicit definition of strategies, which she defines as "an approach that is not domain specific and is of general character, which focuses on the goal and the task as a whole, but which is flexible enough to allow for several ways to proceed when solving a problem" (Fülöp, 2015, p. 49). Fülöp (2015) adds that strategies can be considered as abstractions and as inventions, since they are subjective and imaginary.

In their work about how children use strategies, Siegler and Jenkins (1989) give a similar definition of strategy as "any procedure that is nonobligatory and goal directed" (Siegler & Jenkins, 1989, p. 11) and "as relatively grand entities that encompass a variety of means toward an end" (Siegler & Jenkins, 1999, p. 11). According to Siegler and Jenkins (1989), strategies consist of both conscious or unconscious activities, whereas in comparison, a consciously adopted strategy is called *plan*, which is a "behavior that is voluntary, self-conscious, and intended" (Scholnick & Friedman, 1987, p. 5 as cited in Siegler & Jenkins, 1989, p. 12). Regarding the activation of strategies, different strategies can be competing against each other for activation, if the goal has not been reached yet (cf. Siegler & Jenkins, 1989).

Based on these theoretical assumptions, the term *strategy* is used in this present study to denote an action which is needed by a subject to reach the goal in a particular task, and therefore the development and choice of strategies is influenced by the nature of the task (cf. Plath, 2014) and by the subject's knowledge and experience (cf. Rott, 2011). Based on Siegler and Jenkins (1989)'s assumption, an action should then be denoted as strategy, when there is the possibility of choice, otherwise it should be regarded as a demand of the task. In the next section, different types of strategies used in solving spatial tasks from previous litertaure are introduced and discussed.

2.1.5.2 Holistic and analytic strategies

A well-known differentiation in strategies of solving spatial tasks is Barrat's (1953) difference between whole (holistic strategies) and part approach (analytic strategies) in mental rotation tasks. Holistic strategies or *spatial manipulation* (cf. Burin, Delgado & Prieto, 2000), are strategies in which learners mentally transform or manipulate the whole object or its parts or move around the object in order to solve the spatial task (cf. Barrat, 1953). In analytic strategies, also known as *feature comparison* (cf. Burin et al., 2000), learners compare details of the orginal object and then decide whether the original and the comparing object are identical (cf. Barratt, 1953). Analytic strategies tend to require the learner to focus on specific characteristics of the object, describe these verbally analytic or use logical reasoning to solve the underlying spatial task (cf. Barratt, 1953).

The following phrases are examples of students' verbal thinking in solving a spatial task using holistic or analytic strategy respectively:

1) "I'd try to put figures on top of one another and twist them around" (Barratt, 1953, p. 22).

2) "What I did was orient myself by some identifying mark as the curved line at the bottom; then I noticed another marking symbol as either going to the left or right when I had it in this position" (Barratt, 1953, p. 22).

In the first phrase, the student describes the movement of the figure(s), which indicates the use of a holistic strategy. In the second one, the student identifies symbols or marks, and compares the details when the figure is in different spatial positions. Moreover, the use of holistic strategies (as in phrase (1)) shows that a verbalisation of thinking is still possible but to a limited extent, since the student uses spatial imagery mapping during the movement to solve the spatial task, rather than particular characteristics which could be verbalised more easily.

Some researchers (e.g., Maier, 1999; Plath, 2014) seem to agree that the nature of the spatial task influences the strategy choice among analytic and holistic. For

25

instance, as Maier (1999) points out in his spatial ability model (see Section 2.1.3.3), analytic strategies are expected to be used more in spatial tasks requiring the spatial components *visualization* or *spatial relations*. Spatial tasks assigned to *mental rotation* or *spatial orientation* should activate holistic strategies for their solving (cf. Maier, 1999) (see Figure 1). Other researchers (e.g., Grüßing, 2002) argument that the strategy choice between analytic and holistic is influenced by the students' mathematics performance. In her study about students solving spatial tasks, Grüßing (2002) concludes that there is a tendency that high-performance students use analytic strategies more efficiently than low-performance students (cf. Grüßing, 2002). In other studies about strategic thinking in general, researchers (e.g., Schwank, 2003) state that the choice of strategies is highly dependent on the subject's preferences, rather than solely on the nature of task.

Another differentiation between two types of strategies in solving spatial tasks, *three-dimensional* and *two-dimensional*, are introduced in the subsequent subsection.

2.1.5.3 Three-dimensional and two-dimensional strategies

Three-dimensional and *two-dimensional* strategies are two other strategies which were identified during the solving of written spatial tasks in Gittler's (1984) study about how learners solve three-dimensional cube tests. In *three-dimensional* strategies, learners solve the underlying spatial task by developing three-dimensional mental models, which are used, transformed, moved etc., for solving the spatial task (cf. Gittler, 1984). In contrast, learners using *two-dimensional* strategies produce two dimensional images of the three-dimensional spatial object to facilitate the solving of the spatial task (cf. Gittler, 1984), hence neglecting the third dimension in space. For an example of three-dimensional and two-dimensional strategies, consider an item from the *3D Cube Test* from Gittler (1984), a test for measuring spatial abilities whereby students must control whether one of the given six cubes corresponds to the original cube (see Figure 3).

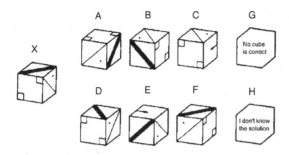

Figure 3   3D Cube test item from Gittler's spatial ability assessment test (cf. Gittler, 1984, p.144)[9]

In the test item in Figure 3, the students are required to find out which cubes A to C correspond to the left most cube. A student who applies three-dimensional strategies is expected to develop a three-dimesional mental image of the cube and mentally rotate it and compare the spatial relations between the different patterns on the surfaces. In contrast, students who use two-dimensional strategies would rather perceive the cube as two-dimensional, as depicted in the test item on the paper (see Figure 3), and neglect the existence of the three not directly visible cube surfaces in the background. Therefore, in this case students would rotate the cube as a figure constructed from the three visible parts (which can be perceived as three adjacent parallelograms in Figure 3), and propose Cube A in Figure 3 as a solution. In his study, Gittler (1984) concludes that the majority of the test persons do not create three dimensional images of the three-dimensional objects and manipulate these, but instead they tend to use two-dimensional thinking to solve the spatial task.

With regard to the previous differentiation between analytic and holistic strategies in Section 2.1.5.2, three-dimensional and two-dimensional strategies can be described using holistic and analytic strategy groups respectively. In three-dimensional thinking, students develop a mental image of the underlying spatial object wholly, which coincides with the holistic approach to solving spatial tasks. In comparison, students applying analytic strategies focus on details of the object, in the case of two-dimensional strategies, on part of the three-dimensional object. For instance, students applying holistic strategies in the task

---

[9]   The original name of the test in German is *Dreidimensionaler Würfeltest* and it was translated by A. M.

27

visualised in Figure 3 manipulate a three-dimensional model of the corresponding cube, whilst students using an analytic approach develop a two-dimensional model and rely more on the properties, such as the spatial relations between the three-surfaces and the patterns on them.

2.1.6 Sex differences in spatial ability performance

There is a considerable amount of literature on students' spatial abilities which shows evidence that there are differences between male and female participants when solving spatial tasks (e.g., Rost, 1977; Linn & Petersen, 1985; Geiser, Lehmann and Quaiser-Pohl, 2006; Büchter, 2011). In their meta analysis study about test persons solving different spatial tasks (different tasks for different spatial components according to their spatial ability model, see Section 2.1.3.2), Linn & Petersen (1985) show that there are sex differences when it comes to performance in tasks requiring spatial abilities. Following an analysis of the achieved results under consideration of sex, it was concluded that male test persons' performance was significantly better in tests of spatial perception and mental rotation than that of the female test persons (cf. Linn & Petersen, 1985). No sex differences were found in the comparison of males' and females' results of the spatial visualization tests.

In his research about sex differences in mathematics performance and spatial abilities, Büchter (2011) shows that the sex differences in mathematics performance can be statistically completely accounted for by sex differences in spatial abilities. In particular, Büchter (2011) states that the component of mental rotation in spatial abilities plays a major role in highlighting the sex differences. Similarly, other studies, like that of Masters and Sanders (1993), confirm the sex differences in spatial ability performance when learners solve tasks regarding the spatial component of mental rotation.

Other researchers (e.g., Rost, 1977) investigated students' spatial abilities and verbal competencies in their quantitive studies. In his research about sex differences in solving spatial tasks, Rost (1977) concludes that whilst female participants show better performance in verbalisations, the male counterparts achieve better results in tasks requiring more logical reasoning and spatial abilities including spatial orientation. Burton, Henninger and Hafetz (2005) conclude similar results in their quantitative research about sex differences and evaluation of the relations between spatial abilities, verbal fluency and finger-length ratios.

Whereas the male participants in their study performed better in tasks demanding mental rotation, female participants performed better in verbal fluency (cf. Burton et al., 2005). Similarly, Geiser, Lehmann and Quaiser-Pohl (2006) point out that there are sex differences concerning verbal and spatial abilities: "whereas females outperform males on measures of verbal fluency, males outperform females on certain tests of spatial ability" (Geiser et al., 2006, p. 683).

In their study about the relationship between verbal and spatial processes, Lanca and Kirby (1995) investigated students' memory and learning of contour maps in four different groups: a verbal learning group, a spatial learning group, a combined verbal and spatial group and a study-only control group. The results of their study indicate better male performance in learning spatial instructions and in three-dimensional map information. Female students in non-spatial task groups tended to memorise two dimensional maps better than in spatial task groups, however, no effect of spatial abilities could be found in the case of map memory (c.f. Lanca & Kirby, 1995). Sex differences in spatial tasks were also identified in the study of Coluccia, Iosue and Bradimonte (2007) about the relationship between map drawing skills and spatial orientation abilities. Coluccia et al. (2007) suggest that males use other strategies for learning maps than females. In particular, males tended to make more use of a holistic approach, whereas females tend to focus more on local features, thus preferring analytic approaches (cf. Coluccia et al., 2007).

However, there are studies, such as the one from Caldera et al. (1999), which do not show any significant differences between male and female participants when solving spatial tasks. In their study about strategies in spatial visualization tasks, Burin, Delgado and Pieto (2000) also conclude that sex does not play a significant role when students use strategies to solve spatial tasks.

The fact that a considerable high number of the studies show that there are sex differences in performance of solving spatial tasks, leads to the questioning about an adequate explanation for this phenomenon. Bergvall, Sorby and Worthen (1994) explain that genetical and environmental factors are responsible for the sex differences in spatial abilities. Whilst Linn and Petersen (1985) reject the genetic reason for such sex differences, the assumption that environmental factors are responsible for the sex differences is emphasised in some empirical studies about sex differences in spatial abilities (e.g., Vasta, Knott & Gaze, 1996). Vasta, Knott and Gaze (1996) state that the fact that spatial abilities can be trained by suitable tasks and experiences indicate that the subject's environment could play a more important role than sex (cf. Vasta, Knott & Gaze, 1996). However, the deviation of two strands in research of spatial ability, the strand which argues that sex does play an important role in solving spatial tasks, and the other strand which denies this claim, show that disagreement and contradications in this area of research. However, since a substantial amount of literature indicates possible sex differences in solving of spatial tasks, which mostly concerns the test items of mental rotation or spatial orientation (e.g., Linn & Petersen, 1985; Coluccia et al., 2007, Büchter, 2011), it is assumed that sex may play an important role in solving spatial tasks. Some studies (e.g., Geiser et al., 2006; Burton et al., 2010) have researched both spatial and verbal competenices of studies in solving spatial tasks quantitatively. Most of these studies show the tendency that the female participants show better verbal abilities, whereas the male counterparts have performed better in spatial tasks, especially in tasks requiring mental rotation or spatial orientation. Such findings show that students' sex as an impacting factor is worth further investigation when analysing how students solve spatial tasks and their use of language in the task solving process.

2.1.7 Summary: Spatial ability

Spatial ability entails different abilities depending on the approach, whether psychological or mathematics educational, and on the intentions, such as analysis of internal processes or learning and task solving processes. However, researchers (e.g., Maier 1999; Plath, 2014) seem to agree that visual perception, construction of mental images, processing information, and thinking are cognitive processes which are characteristic for solving spatial ability tasks.

From a mathematics educational perspective, I believe that both mathematics educational and psychological models of spatial abilities are important for investigating the notion of spatial abilities. Whereas the models in psychology offer a way to assess students' spatial ability performance and structure the important components which constitute spatial ability, models in mathematics education offer perspectives to integrate spatial abilities in learning and teaching of mathematics, such as the verbalisation of properties and construction of solids. The consideration of Linn and Petersen's (1985) model for assessment of students' spatial ability performance and Pinkernell's (2003)'s model for spatial task design plays an important role in the theoretical framework about spatial abilities in this present study. In this study, the notion of spatial abilities is considered as a collection of abilities, which encompasses not only mental abilities (i.e. spatial thinking), such as mental manipulation of spatial objects, but also spatial-motoric skills, such as performing real actions on building cubes, and spatial-linguisic skills, such as describing a spatial object.

Following the definition of the notion of strategy as a goal-oriented procedure and flexible regarding its use in solving tasks, an overview of the most common strategies for solving spatial tasks from previous literature have been discussed in Section 2.1.5. It was agreed upon by researchers that the use of strategies is highly dependent on the subject's thinking and experiences, rather than only on the nature of the spatial task. However, the subject's sex could also play a role in solving spatial tasks, as numerous previous studies about sex differences in spatial abilities have shown (see Section 2.1.6). Findings from previous research on spatial abilities (e.g., Geiser et al., 2006; Burton et al., 2010) show that males outperform females when it comes to solving spatial tasks linked to mental rotation or spatial orientation, whereas females are better at verbal competencies in general. The latter competencies are addressed in general in the next section about the use of language in learning and teaching mathematics.

2.2 Language in mathematics classroom

The central role of language in learning and teaching mathematics has been increasingly recognised in the field of mathematics education. Research about language in mathematics education has pointed out the complex relationship between language and mathematics education, which has been summarised by Durkin (1991) in the following phrases:

> Mathematics education begins and proceeds in language, it advances and stumbles because of language, and its outcomes are often assessed in language. Such observations could be made of most school curricula, but the interweaving of mathematics and language is particularly intricate and intriguing. (Durkin, 1991, p. 3).

This section provides an overview of the interweaving of mathematics and language based on results from previous literature. After giving a definition of language from several disciplines, the relationship between language and thinking based on Vygotski's (1934/1993) theoretical framework are discussed. This is followed by the description of two important functions of language in mathematics learning and teaching.

Regarding the use of language in mathematics classroom, there are different varieties of language spoken by students and teachers in mathematics, which is described in Section 2.2.4. Section 2.2.5 focuses on the use of metaphors for learning and teaching mathematics. This is followed by a discussion of the importance of language in German mathematics curricula in the last section of this sub-chapter.

2.2.1 Definition of the notion of language

Language is one of the most important aspects in learning and teaching of knowledge and is a phenomenon which accompanies us throughout everyday life to communicate and understand each other and to represent objects and situations in the world. The linguist Ferdinand de Saussure (1916) differentiates between three aspects of language: *langage,* which refers to the abstract concept of human language as a biological propert; *langue,* which denotes a specific language system and their set of rules for generating *langage,* for example, the German or English language; and *parole,* which denotes the concrete use of speech in a particular language system (cf. de Saussure, 1916)[10]. However, due to the global use of language in different disciplines there are various definitions of and approaches to the notion of language. In structuralism and formal theories of grammar, language can be considered as a system of signs determined by syntactical rules to generate words and sentences for communication, whereby these syntactical rules are a finite set and language specific (cf. Chomsky, 1995). From a cognitive and neuropsychologist perspective, language compromises the mental processes in the human brain significant for language acquisition and development and understanding of language (cf. Lesser, 1989). One of the basic assumptions of this view of language is the innateness of language, which means that language is produced in the brain rather than learnt through experience (cf. Lesser, 1989).

The communicative aspect of language plays an important role in sociolinguistics and pragmatics. In this domain, language is defined as a communication system consisting of signs and meanings used by human beings to express themselves and to communicate with other human beings (cf. Lyons, 1981). In con-

---

[10] In this thesis, the words *parole* and *spoken language* are used interchangeably. The words *langue* and *language* should retain the same meaning as in de Saussure's (1916) theoretical framework.

trast to formal theories of grammar language, in sociolinguistics, the rules of language are considered to be the result of the process of communication among its speakers, thus communication is a major catalyst for the development of syntactic rules of language (cf. van Valin, 2001). The importance of communication in social context is embedded in the notion of discourse, which Bruner (1966/1971) describes as:

> Discourse consists, in its essentials, of an *addresser*, and an *addressee*, a contact that joins them, a message passing between them, a context in which the message refers and a linguistic code that governs the way in which messages are put together and things referred to. (Bruner, 1971, p. 106)

In this present study, the sociolinguistic view of the concept of language is the most adequate, because language is vital for communication and as developing in a social environment through experience and interaction with other individuals. Although syntactical structures of language are important for developing appropriate communication, these should not be regarded as the only important units to construct language, because one should not neglect acceptability, which is a feature often observed in speech and which does not hinder communication. For instance, consider the phrase *a giraffe is a type of plant,* which is grammatically correct, but not considered as appropriate in speech due to its underlying meaning, hence not acceptable. Therefore, the aspect of acceptability shows how social interaction influences the use and meaning of language, which justify Evans and Levinson's (2009) claim that many aspects of language are shaped by communication in a social context.

Other concepts of language which are important for the present study are language proficiency and language acquisition. Language proficiency denotes a speaker's language skills or competencies, which are the ability to speak, read, write, listen and understand a particular language, which can be the first, second or a foreign language. Moreover, language acquisition describes the learning process of understanding a language and of communicating in the same language by generating words and sentences. This process requires learners to develop phonetic, morphological, syntactical and semantical features of a language. Therefore, language proficiency can be regarded as an indicator for a students' language acquisition. In the context of language use in education, language and

33

mathematics education researchers differentiate between everyday language and academic language in language acquisition in classrooms. A more detailed description of these levels of acquisition in mathematics education is discussed in Section 2.2.4.

2.2.2 Language and thinking

Language plays an important role in the development of thinking, since language is a medium for generalising and abstracting concepts, which are used to organise experiences, and act as mediator between the subject and the object of thought (cf. Vygotsky, 1993). Lew Vygotsky was one of the first psychologists who investigated the relationship between language and thinking extensively and systematically using a psychological experimental approach. Among his works, *Thought and Language* provides an extensive description of the inner connection between speech and cognitive concepts. Vygotsky (1934/1993) describes the relationship between thought and language as a complex dynamic process characterised by movement from thought to word and vice versa. However, in this process, thought is not just expressed via language, but emerges through the use of words (cf. Vygostky, 1993). The connection between thought and language leads to the analysis of meanings of words, which is described as the union of word and thought:

> Word meaning is an elementary 'cell' that cannot be further analyzed and represents the most elementary form of the unity between thought and word. The meaning of a word represents such a close amalgram of thought and language that it is hard to tell whether it is a phenomenon of speech or a phenomenon of thought (…). It is a phenomenon of verbal thought, or meaningful speech – a union of word and thought (Vygotsky, 1993, p. 212)

Hence, the discussion of the relationship between language and thinking demands a deeper investigation of word meanings. Vygostky (1934/1993) points out the need to differentiate between two types of speech, internal and external speech, for a deeper understanding of word meaning. [11] The first type of speech

---

[11]   The notion of language in Vygostky's (1934) work seems to concide with the notion of *langage* in de Saussure's (1916) work. However, in contrast to de Saussure's (1916) notion of *parole*, which denotes the external speech in concrete terms, Vygostky includes the mental verbal representations in the cognitive system in his notion of speech.

inner speech is required by the use of language and which serves as an instrument of thinking. *Inner speech* denotes the semantics of speech and can be considered as a truncated form of the external speech in which the addressee is oneself. One of the characteristics of inner speech observed in the Vygotsky's (1934/1993) experiments is its syntactical fragmentation when compared to the external speech. In particular, during the observation of students' inner speech, which was carried out by an analysis of students' speech addressed to themselves (i.e. egocentric speech according to Piaget (1951)), a tendency of subject and object omission, whereas students reduce an expression to its predicate, could be noticed (cf. Vygostky, 1993). To investigate the area of inner speech more deeply, Vygotsky (1934/1993) states that inner speech develops from egocentric speech, a concept developed from Piaget (1951) and which Vygotsky (1993) considers as a transitional process from a child's social collective activity to a more invidualised one. Egocentric speech is based on the concept of a child's egocentrism, in which children between the ages of three and five focus on themselves and think that other human beings have their same thoughts, feelings and experiences (cf. Vygotksy, 1993). Vygotsky (1993) considers egocentric speech as an essential part of the social learning process and considers it as means to examine childrens' inner speech by using experimentation and observation to externalise inner speech. This process of externalisation of inner speech can be facilitated by the design and implementation of an outer activity in which the child is encouraged to engage in social speech (i.e. speech addressed to others rather than oneself) which would enable the objective functional analysis of inner speech (cf. Vygotsky, 1993). An important criterion for the existence of social speech is the feeling of being understood, which is only present if a social situation can be created during the experiment (cf. Vygotsky, 1993).

The second type of speech, *external speech*, is defined as speech generated for other addressees rather than oneself and therefore focuses more on the phonetic aspect of speech (cf. Vygostky, 1993). Vygostky (1993) argues that it is wrong to consider the inner speech as being existent before the external speech or as being the reproduction of external speech in memory. However, he suggests that external and inner speech can be considered as opposites having entirely separate speech functions (cf. Vygotsky, 1993)). Although abbreviation can also be observed in external language, Vygostky (1934/1993) describes this phenomenon as rule for internal speech and as an exception for external speech. Another difference between *inner* and *external speech* is the agglutination of words in the former speech to combine them, which is also a typical characteristic of a child' egocentric speech (cf. Vygotsky, 1993).

Vygotsky (1934/1993) emphasises that inner speech is a speech which should not be considered merely as an interior aspect of external speech. He summarises the external speech as the verbalisation of thoughts into words and thus materialising them, whilst internal speech transforms speech into thoughts (cf. Vygotsky, 1993). However, one should not consider the existence of the corresponding units in both speech and thoughts and there are no separate units in thinking as in speech: "The two processes are not identical, and there is no rigid correspondence between the units of thought and speech" (Vygotsky, 1993, p. 249). Consider, for instance, a situation in which a child is playing football – this corresponds to one thought, but several units of speech, for example, *child*, *play* and *football*.

A possible way to describe the process of verbal thought or verbalised thinking is as a development of thought which is triggered by a motive, shaped in the inner speech and in word meanings and ends in words or external speech (cf. Vygostky, 1993). Nevertheless, the relationship between language and thinking remains complex, since the process of verbal thought might involve movements back and forth along the different phases (cf. Vygotsky, 1993). However, a simplification of the verbal thought sheds light on the complex relationship between language and thinking. The investigation of this binary interrelationship shows that the understanding of language requires the understanding of thoughts. Vice versa, the understanding of language is not sufficient without deciphering and understanding of the underlying thoughts, because most often there is always an intention for the use of language. This leads us to the two functions of language in the mathematics educational context, which are described in the next section.

2.2.3 The functions of language

Language plays an important role in mathematics learning, just as in every learning process. In previous research about language use in mathematics classroom (e.g., Maier & Schweiger, 1999; Wessel, 2015) two different functions of language – the communicative function and the cognitive function of language – were established. The former denotes the use of language for understanding each other, by focusing on communicating one's thoughts, views and considerations to other persons (cf. Maier & Schweiger, 1999).

Language in its cognitive function is used to acquire new knowledge through the mediation of information with the help of conceptual representations (cf. Maier & Schweiger, 1999). Elements of language and conceptual learning induce the construction of new knowledge and enable or facilitate its activation in the cognitive system (cf. Maier & Schweiger, 1999). Wessel (2015) points out that the background of the cognitive function of language is based on the assumption that language is a cognitive tool which facilitates thinking and understanding processes. Consider the following task for a deeper understanding of the communicative and the cognitive functions of language: *"draw a square, a rectangle, and a quadrilateral on a piece of paper and write down the characteristics of each figure"*. On a communicative level, the function of this task is to communicate that the student has to solve this particular task which can be conveyed by spoken or written form of language. On a cognitive level, the students learn to understand the different notions *square, rectangle* and *quadrilateral* and differentiate between them in order to derive their different characteristics by using language and activating and manipulating verbal information in their cognitive system.

Both functions of language do overlap and should not be considered as separate entities (cf. Wessel, 2015). Maier and Schweiger (1999) describe the communicative role of language as being an amplifying factor on the cognitive function, because the aim of the former is to support the goal of the cognitive function. This supports the claim that communication is vital for the construction of cognitive processes, which can be seen as a joint approach from the sociolinguistic and cognitive perspective of the concept of language.

2.2.4 Levels of language acquisition

As already mentioned in Section 2.2.1, language acquisition is the process of acquiring the different aspects of a particular language. The process of language acquisition does not only refer to the acquisition of a new foreign language, or of the mother tongue during childhood, but it also includes the learning of new registers within the language which an individual can speak fluently. In sociolinguistics, a register can be defined as a variety of language[12] which is used in a

---

[12]   In this thesis, the term *variety of language* denotes a language with different characteristics (mostly on a semantic or syntactical level) within *langue* (cf. de Saussure, 1916) deriving from its use in different contexts, in this case, in mathematics teaching and

particular environment for a particular purpose. An example of register is legal English or legal German, which does not only include specific vocabulary, but also complex sentence structure and other linguistic aspects typical in such registers.

Different registers or varieties of language can also be observed in the different languages used for learning and teaching in classrooms:

> The talk of teachers and students draws together – or bridges – the 'everyday' language of students learning through English as a second language, and the language associated with the academic registers of school which they must learn to control. (Gibbons, 2006, p. 1)

A popular theoretical framework about the use of language for different purposes in educational context is Cummins' (1979) differentiation between two levels of language acquisition: *Basic Interpersonal Communication Skills* (BICS) and *Cognitive Academic Language Proficiency* (CALP). BICS denotes the abilities to communicate in a language in everyday-life situations with other speakers of the same language. This type of language is characterised by the use of non-verbal language (such as gestures) to facilitate communication and understanding (cf. Cummins, 1979). Moreover, social commucation in BICS tends to be context-embedded, which means that a common context is required for the speakers to understand each other. Such a language variety tends to be used to communicate with friends and relatives and does not require complex syntactical structures or specific vocabulary which is not used in everyday life situations (cf. Cummins, 1979).

In contrast, CALP can be understood as an extension of BICS, because it denotes the ability to use oral and written language to additionally understand academic concepts and ideas. This level of language acquisition is characterised by the extensive use of complex syntactical structures and academic vocabulary (cf. Cummins, 1979). The vocabulary used is influenced by the topic or subject of the conversation or situation, for example, mathematics, physics, biology, chemistry, etc. This supports the idea that the level of proficiency in CALP has a ma-

---

learning. This differs from the traditional association of the term varieties of language in linguistics, which tends to include dialects, sociolects and other varieties spoken by cultural groups.

jor impact on students' performance in school. In contrast to BICS, a context is not necessary in CALP and non-verbal language is not as important as in BICS:

CALP is said to occur in 'context-reduced' academic situations. Where high order thinking skills (e.g., analysis, synthesis, evaluation) are required in Curriculum, language is 'disembedded' from a meaningful, supportive context (Baker, 2006, p. 174).

In mathematics educational context, CALP can be translated to mathematics language which is the language commonly used in mathematics textbooks, in teachers' or mathematics education researchers' language or in academic texts by mathematics educators.[13] Halliday (1978) describes mathematics language as "the meaning that belong to the language of mathematics (the mathematical use of natural language, that is: not mathematics itself), and that a language must express if it is being used for mathematical purposes" (Halliday, 1978, p. 195). However, due to the different areas of mathematics mathematics language should be considered as a generalised term to denote the collection of different sub-registers used in mathematics classroom, for example, spatial sub-register in geometry classes, functional sub-register in calculus, etc.

Cummin's (1979) distinction between BICS or everyday language and CALP or mathematics language has served as a basis for the investigation of different varieties of language in learning and teaching context in mathematics classroom. However, both varieties of language should not be considered as entirely separate entities, but rather as a continuum (cf. Gibbons, 2006; Clarkson, 2009; Prediger & Wessel, 2011), in which the degree of abstraction increases (from everyday to mathematics language) or decreases (from mathematics to everyday language). An assumption which supports the representation of everyday and mathematics language as a continuum is the fact that a solid knowledge of everyday language serves as fundamental basis for the acquisition of academic language, in particular mathematics language. However, as Leisen (2011) points out, there are other varieties of language, such as instruction language, i.e. the language

---

[13] In models about levels of language acquisition in mathematics education, e.g. from Prediger and Wessel (2011), researchers differ between three levels of language acquisition: everyday language (*Alltagssprache*), academic language (*Bildungssprache*) and technical language (e.g., mathematics language) (*Fachsprache*), the last two being more closely related in terms of their properties. In this work, CALP in mathematics education shall refer to the language used for expressing mathematical ideas in mathematics classroom, hence, it will be referred to as mathematics language.

used in learning and teaching in educational context, regardless of the subject. Other language varieties include visual language, which serves for visualising and explaining knowledge and procedures using graphs, pictures, diagrams, etc., and symbolic or formal language, which is used to represent knowledge and procedures using symbols, formulas, etc. (cf. Leisen, 2011). The latter varieties of language show how different representations of mathematical knowledge (see Section 2.4.1) influence the kind of language used in mathematics classroom.

2.2.5 Metaphors in mathematics classroom

The use of metaphors in learning mathematics was pointed out by different studies about language in mathematics classroom (e.g., Pimm, 1981; Sfard, 1998; Lakoff & Núñez, 2000; Malle, 2009; Font, Godino, Planas & Acevedo, 2010). According to Lakoff and Johnson (1980), metaphors are devices used for exploration and understanding of a particular idea through another one. A similar description of metaphors can be found in linguistics, whereby it is defined as a rhetorical device that transfers a meaning from one word to another. However, metaphors should not only be considered just as figures of speech or an embellishment in a discourse, but rather as medium to enable and support thinking about abstract ideas (cf. Lakoff & Núñez, 2000). Sfard (1998) states that metaphors are important objects of analysis in learning mathematics due to following reasons:

> Their [metaphors] special power stems from the fact that they often cross the borders between the spontaneous and the scientific, between the intuitive and the formal. Conveyed through language from one domain to another, they enable conceptual osmosis between everyday and scientific discourses, letting our primary intuitition shape scientific ideas and the formal conceptions feed back into the intuition. (Sfard, 1998, p. 4)

Mathematics and mathematical discourse is based on the use of metaphors, which Malle (2009) describes as words or phrases with a 'transferred' meaning, i.e. the meaning originates from another non-mathematical field (cf. Malle, 2009). An example of a metaphor in mathematics education is the phrase "an isosceles triangle is like a human standing on two feet, therefore both sides are equal" (Malle, 2009, p. 10). Malle (2009) claims that most of the mathematical terms originate from everyday language (for example, *tree, root, area, divide, break-down, real, similar, proportion,* etc.) and other ones, like *function, integral, differentiate, prime, trigonometry, diagonal,* etc. also originate from everyday language of either Latin or Greek. However, sometimes the meaning has only been partially transferred from everyday to mathematics language, as in the case of the meaning of the mathematical term *root* from the Latin word *radix,* which means base, foundation or origin (cf. Malle, 2009).

Based on Lakoff & Johnson's (1980) previous work, Lakoff & Núñez (2000) define conceptual metaphors as mechanisms, which support the understanding of abstract ideas through concrete terms. Metaphors are perceived to be projections from source domains to target domains, in which the sources' properties and characteristics are assigned to the target (cf. Lakoff & Núñez, 2000; Font et al., 2010). There are two types of conceptual metaphors used in mathematics: *grounding* and *linking* metaphors (cf. Lakoff & Núñez, 2000). Whilst grounding metaphors are metaphors which project a source outside the field of mathematics to a target within mathematics, in linking metaphors both source and target orginate within mathematics, but tend to be anchored in different fields of mathematics (cf. Font et al., 2010). An example of a grounding metaphor is the phrase, "classes are containers" (Núñez, 2000, p. 13). In this example, properties of the source *container* are projected to the notion of class (see Table 2).

Source domain	Target domain
Container schemas	Classes
Interiors of container schema	Classes
Objects in an interior	Class members
Being an object in an interior	The membership relation
An interior of one container schema	A subclass in a larger class

Table 2   The grounding metaphor *classes are containers* (Núñez, 2000, p. 13)

The linking between contexts and situations in everyday life and abstract ideas is a fundamental principle for learning of mathematics and constructing abstract mathematical knowledge in the classroom. Therefore, metaphors are often used to construct and reconstruct mathematical knowledge, especially when considering that "(…) a larger number of the most basic, as well as the most sophisticated, mathematical ideas are metaphorical in nature" (Lakoff & Núñez, 2000, p. 364). Metaphors are important tools to facilitate mathematics learning, since the desirable meaning of mathematical terms in mathematics language can be transferred from everyday language (cf. Malle, 2009). However, a heavy reliance on such a meaning transfer to learn mathematics can also lead to barriers in teaching mathematics (cf. Malle, 2009). For instance, most students (especially at the beginning of secondary level) tend to understand the mathematical concept of *diagonal* as *oblique* or *slanting*, as in the case of use of the word *diagonal* in eve-

41

ryday language, sometimes also in geometry classes (for example, in typical examples of rectangles and parallelograms) (cf. Malle, 2009). Whereas the use of such a metaphor is helpful, it might also be hindering, since not every slanting line is a diagonal (cf. Malle, 2009). Several mathematics education researchers (e.g., Sfard, 1998; Malle, 2009) have pointed out how metaphors can act as obstacles in mathematics learning and teaching. However, most researchers in mathematics education (e.g., Pimm, 1981; Lakoff & Núñez, 2000; Malle 2009) seem to agree that metaphors are important elements in mathematical discourse to learn and teach mathematical knowledge, and for supporting the understanding of abstract mathematical ideas. This present study shows the importance of metaphors in mathematical discourse and illustrates the limitations of metaphors when communicating and constructing mathematical-spatial knowledge.

2.2.6 Roles of language in German mathematics curriculum and classroom

The three main roles of language in mathematics education upon which researchers (e.g., Prediger, 2013; Wessel, 2015; Schütte, 2009; Knapp, 2006) seem to agree are: language as a learning medium, language as a learning goal or language as a barrier in learning and teaching of mathematics (cf. Prediger, 2013).

Language as a learning medium means that language is used to represent and communicate content knowledge via linguistic elements (cf. Knapp, 2006). Wessel (2015) states that the student language acts as a learning medium in the process of learning content. She emphasises the didactical principles of Wagenschein (1989) and Winter (1996), who argue that students should be competent in their mother tongue and then in the respective academic language and that knowledge should be first represented in everyday language and then in the academic language respectively (cf. Wessel, 2015). The emphasis on language as a learning medium is also documented in German mathematics curriculum as a goal of mathematics classes:

> Cognitive processes related to content knowledge, concept formation, and the assessment and evaluation of mathematical issues and problems are mediated via language, as well as the presentation of learning results and its discussion in communication. (North-Rhine Westphalia Ministry of Education, 2011, p. 10)[14]

---

[14]   This section was translated by A. M. and is reproduced without any further alterations from the curriculum documents of North-Rhine Westphalia Ministry of Education (2011).

However, the teacher's mathematical language or language used in textbooks can also act as a learning medium. For instance, tasks in current mathematics textbooks do not only provide calculations, but also provide or demand descriptions or explanations. To solve or understand tasks with descriptions or explanations learners are not only required to understand the mathematical language but also to produce language.

Language as a learning goal is based on the belief that the aim of mathematics lessons should not only be the learning of content knowledge, but also of linguistic knowledge associated to language used in mathematics classrooms (cf. Prediger, 2013). Language learning has been gaining importance in mathematics lessons, as the following excerpt from the German mathematics curriculum shows:

> The acquisition of mathematics content learning is linked intensively to the development of language skills. (...). Such language skills [e.g., presentation of learning results, communication etc.] do not develop just from the everyday language competencies, but they should be initiated and further developed in language-sensitive mathematics lessons. (North-Rhine Westphalia Ministry of Education, 2011, p.10)[15]

There are several specific goals which teachers need to consider in order to develop a language-sensitive mathematics lesson. For instance, teachers should ensure that secondary students describe phenomena by using mathematics language, that they use the right tenses when describing results, processes or algorithms and that they use adequate and complex sentence structure in order to represent relations (cf. North-Rhine Westphalia Ministry of Education, 2011). Another focus on learning of language skills in German mathematics curriculum can be observed in the high importance of argumentation and communication for mathematics learning (cf. Federal German Ministry of Education, 2004). In particular, students at the end of lower secondary level are required to communicate mathematical facts comprehensible, and to be able to justify them (cf. North-Rhine Westphalia Ministry of Education, 2014). Further students' abilities at this level, formulated as reasoning and communication skills in mathematics curriculums in North-Rhine Westphalia, include reading skills, such as the ability to extract, structure and evaluate mathematical information from texts, pictures and

---

[15] This section was translated by A. M. and is reproduced without any further alterations from the curriculum documents of North-Rhine Westphalia Ministry of Education (2011).

tables, communication skills, such as the ability to communicate and present mathematical problem solving processes in one's own words and to use mathematical terms, and other cognitive-linguistic skills, such as the ability to construct concept networks and relation networking (cf. North-Rhine Westphalia Ministry of Education, 2014). Therefore, students are not only expected to learn new vocabulary in mathematics, but also to learn how to communicate, argument and reason, which are also important aspects of language learning.

However, language can also act as a barrier in learning processes in mathematics learning and teaching (cf. Radford & Barwell, 2016). Clarkson (1992) states that language is a barrier in mathematics learning for multilingual students with low language proficiency in their first language (L1) or second language (L2), who often obtain lower achievements than students with high language proficiency. Therefore, mathematics lessons should also provide language support along with content learning for reducing obstacles in mathematics learning which can arise from linguistic issues.

The above mentioned three roles of language in mathematics learning and teaching show the high influence of language in mathematics lessons, and show the need to integrate language learning and content learning in mathematics education. Whereas the curriculum addresses the need for a higher language awareness and the need for integrative language and content learning in mathematics, the implementation of language and content learning in mathematics lessons is not only influenced by the mathematical topic (e.g., calculus, geometry, probability, etc.), but also by the diverse needs of students with different linguistic backgrounds. Therefore, in mathematics lessons with the focus of language and content learning, teachers need to mediate between the development of different students' language proficiency and the particular mathematical topic.

2.2.7 Summary: Language in mathematics classroom

In Section 2.2, an overview of the theoretical framework about language in mathematics classroom was provided. From a sociolinguistic perspective of language, language is developed and constructed by communication in mathematics lessons between teachers and students. Along with the communicative function of language, language and its cognitive processes is a means for learning and understanding mathematics knowledge (see Section 2.2.3). As it can be observed from Vygotsky's (1934/1993) work, the relationship between thinking and language is rather complicated to describe, but the meaning of language in communication helps to understand better how language and thinking processes are interrelated. Regarding the use of language in mathematics classrooms, there are mainly two different varieties of language or levels of acquisition in mathematics lessons: everyday language and mathematics language from the students' and teachers' perspective respectively (see Section 2.2.4). Both of these levels of language acquisitions make use of metaphors in order to describe abstract mathematical ideas. Based on the theoretical framework of Lakoff and Núñez (2000), metaphors can be considered as projections from a source to a target in mathematics. Depending on whether the metaphor's source orginates outside mathematics or inside mathematics, Lakoff and Núñez (2000) differentiate between grounding and linking metaphors (see Section 2.2.5). Due to the high resemblance of students' language to everyday language, in comparison to teachers' language, which is more a mathematical language, one can hypothesise that students are more likely to use grounding metaphors. The difference between students' and teachers' language can be observed in the need to develop language and content integrated learning in mathematics classroom in particular to support low language proficiency students in their content learning (see Section 2.2.6). However, language and content integrated learning in mathematics classroom is also required for the further development of language skills (regardless of the students' language proficiency) and to raise language-awareness among students.

2.3 Interplay of spatial ability and language

After outlining the current research on spatial ability and on language separately, the relationship between both, which has not been focused on enough in recent mathematics education research, is discussed from a mathematics education perspective. The reason for the negligence of this relationship in research might be the standpoint in traditional psychology in which cognitive processes are likely to be considered as separate entities:

> Interfunctional relations in general have not as yet received the attention they merit. The atomistic functional modes of analysis prevalent during the past decade treated psychic processes in isolation. Methods of research were developed and perfected with a view to studying separate functions, while their interdependence and their organization in the structure of consciousness as a whole remained outside the field of investigation. (Vygotsky, 1934/1993, p. 1)

45

However, the previous theoretical background manifests both spatial ability and language in its cognitive function as cognitive processes in mathematics education. Therefore, this section presents detailed results from previous research about possible intersections between the domains of language and spatial ability. The first section, Section 2.3.1, provides theory on the phenomenon of spatial language from psychology and linguistics research, which is the strongest argument for considering the link between both domains. The next section, Section 2.3.2, presents an attempt to support the connection between both domains by focusing on the role of language in solving spatial tasks.

2.3.1 Spatial language

The phenomenon of spatial language seems to provide a promising direction for research on the relationship between language and spatial abilities. Being more intensively researched in the area of psycholinguistics, some of the most common definitions of spatial language are as *language of objects and places* (cf. Landau & Jackendoff, 1993), *language of space* or *spatial description* (cf. Levinson, 1996; Levinson, 2003; Taylor & Tversky, 1996). The commonality in these definitions is that spatial language is the language used when describing spatial phenomena, whether objects or places. From a mathematics education perspective, spatial language can be defined as the register used in spatial geometry.

In his approach to language in space, Lehmann (2013) uses the notion of *spatial construction* which he defines as the linguistic coding of spatial orientation in a language system. Lehmann's (2013) notion of spatial construction or spatial language emphasises the use of language in order to orientate oneself in space and enriches space by the communicative and cognitive functions of language. However, according to Lehmann (2013), the origin of spatial construction is non-linguistic, because the non-linguistic substrate serves as a universal basis for space construction in different languages worldwide. Moreover, the process of space construction takes place differently depending on the language used (cf. Lehmann, 2013). Similarly, Levinson (1996) points out the importance of language for our development of spatial conceptions by referring to the fact that "many languages do not use the planes through the body to derive spatial coordinates, i.e. they have no left/right/front/back spatial terms" (Levinson, 1996, p. 356). For instance, Australian Aboriginal languages, such as Guugu Yimithirr and Thaayorre, do not have any words for relative directions, such as right or left, but instead use cardinal directions for spatial orientation (cf. Levinson, 1994). Hence, spatial conceptions and spatial language can be regarded as influencing each other to a certain extent.

Cognitive linguists, such as Coventry, Tenbrink and Bateman (2009), indicate the vital importance of spatial language which they describe as enabling us "to find objects and places outside the immediate stimulus reach of speakers, making use of a wealth of spatial relations, spatial perspectives, and spatial objects" (Coventry et al., 2009, p. 1). Moreover, they highlight the importance of spatial language to occur in a dialogue rather than in a monologue, since the interaction in a dialogue gives students the opportunity to participate more actively and creates a less artificial and less restricted setting (cf. Coventry et al., 2009). This kind of setting enables the participants to find common reference frames, identify the required spatial properties and objects and to give feedback to each other to develop and enhance spatial language (cf. Coventry et al., 2009).

Several researchers (e.g., Coventry et al., 2009; Levinson, 1996; Landau & Jackendoff, 1993) emphasise the importance of spatial language to understand the underlying spatial concepts. Levinson (1996) states that an analysis of spatial language is useful to examine the speakers' understanding of spatial concepts (cf. Levinson, 1996). Similarly, Landau and Jackendoff (1993) point out that an appropriate analysis of spatial language enables new insights and knowledge about spatial thinking and the understanding of the corresponding spatial representations. Their concept of spatial representations is defined as "a level of mental representation devoted to encoding the geometric properties of objects in the world and the spatial relationships among them" (Landau & Jackendoff, 1993, p. 217). This definition of spatial representations can be considered as comprising several characteristics of the components of spatial ability in Pinkernell's (2003) model (mental and real spatial visual operations and geometrical thinking) (see Section 2.1.3.4), and of cognitive process in spatial tasks (such as spatial relations perception and mental images) (see Section 2.1.2). Furthermore, Landau & Jackendoff (1993) analyse the relationship between spatial representations and other information and representations, which is visualised in Figure 4. They state that spatial representations, which is used synonymously with spatial information or knowledge, are formed from visual, auditory or haptic information and act as an input for motor information and even for language or linguistic representation. The double arrow between language and spatial representations in Figure 4 shows the reciprocal effect of these two representations, which must exist to account for the phenomenon of spatial language. In order to gain spatial information from language, Landau & Jackendoff (1993) presuppose that "if linguistic judgements can be based on spatial properties of objects then the information involved in the linguistic judgement must be able to pass through the interface between spatial representation and language" (Landau & Jackendoff, 1993, p. 256).

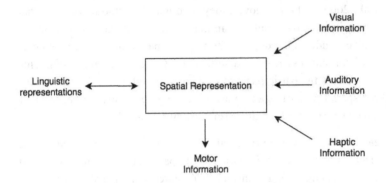

Figure 4   The relationship between spatial representation and language according to Landau and Jackendoff (1993, p. 256).

Therefore, for a better understanding of the relationship between language and spatial ability, one has to analyse deeper the phenomenon of spatial language, which can be simplified as the language used to talk about objects and about places (cf. Landau & Jackendoff, 1993).

2.3.1.1 Functions of spatial language

Based on the theoretical framework of Levinson (2009) and Landau and Jackendoff (1993), there are two main functions of spatial language in discourse: *object description* and *object localization*. The former function of spatial language is to describe objects by referring to their underlying spatial characteristics and representations required for their perception in space. Landau & Jackendoff (1993) state that shape is the most important factor for identifying objects, especially when it comes to choosing names for a particular object, which has not been assigned a common name yet. For the identification of objects based on shape, the spatial representations must provide enough information about the shape so that an individual is able to differentiate between the objects which be-

48

long to a particular category based on the shape (cf. Landau & Jackendoff, 1993). There are several objects which might be described by the same shape, for example, a football and a basketball, however, a more detailed shape description is needed to identify one ball from another. Another useful way to describe objects might be by focussing on the distinct parts of that objects and naming them separately by focussing on their shape and then to agree on a suitable name for the whole object depending on the names of parts (cf. Landau & Jackendoff, 1993). A corresponding example is shown in Figure 5(a), where a particular object is broken down in four parts (represented by the different shadings), each of which can be assigned a particular name, such as *arms*, *head*, etc… and a possible name for the whole object could be a *cube with a head and two arms*.

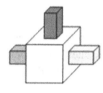

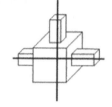

Figure 5(a)   **Breakdown for describing a**   Figure 5(b)   **Oriented axes for describing**
              spatial object   **a spatial object**

A further way to describe spatial objects is to use oriented axes to describe its spatial parts. The oriented axes denote the regions of an object based on spatial orientation, such as the top or bottom of an object, etc. (cf. Landau & Jackendoff, 1993). Figure 5(b) shows an example of how oriented axes can be used to differentiate between the different spatial parts of a particular object. The horizontal axis in Figure 5(b) distinguishes between the top and bottom of the spatial object, whereas the vertical axis serves as left and right spatial orientation. The use of such oriented axes to describe parts of objects can be regarded as spatial orientation within the object, which is described more in the next sub-section, Section 2.3.1.2.

The second function of spatial language is to localise an object in space. Spatial language which is used for object localization is characterised by several linguistic phenomena which represent the cognitive process in spatial tasks such as figure-ground, spatial relation and spatial position perception. Figure-ground perception has been described as the ability to identify a figure or an object from an optical complex background or to identify parts of the object from a given object and isolate them (see Section 2.1.2.1). When describing the location of spatial objects or the spatial relation between two spatial objects, the notions of *figure* and *ground*, are important concepts in order to distinguish between what is localised with regards to what (cf. Talmy, 1983). *Figure* denotes the object which is to be located and is synonym with theme (cf. Carroll, 1988) and reference object or trajectory (cf. Langacker, 1987). Talmy (1983) describes the figure as "a moving or conceptually moveable object whose site, path, or orientation is conceived as variable the particular value of which is the salient issue" (Talmy, 1983, p. 232). In order to locate a figure in space, the locating entity *ground*, also known as *relatum* in Carroll's (1988) work, has to be introduced (cf. Jordens & Lalleman, 1988). In comparison to the figure, the ground is the reference object and is considered to be stationary in the frame of reference. Moreover, the ground's properties should be known in order to characterise the figure's spatial properties (cf. Talmy, 1983). For instance, in the phrase "the book is on the shelf", the book is the figure and the shelf is the ground, whereby the shelf is used to identify the location of the book in space.

The localisation of an object in space can be dependent on the position of the describing person in space, who acts as the centre in spatial discourse (cf. Levinson, 2003, 2006). On a linguistic level, elements in speech which can indicate that the person is the figure or the ground in space are referred to as spatial deixis. Levinson (2006) defines spatial deixis as a linguistic operation in which an individual uses his or her own body or speech to act as a reference point to identify the figure. Spatial deixis are phrases or words which require contextual information bound to a particular spatial situation in order to understand their spatial meaning (cf. Levinson, 2006). According to Levinson (2006), there are three different types of spatial deixis which might be useful when localising an object in space: *central, compositional* deixis and *optional origo*. *Central* deixis are demonstrative pronouns, for example, *this*, or adverbs, for example, *here*, in which the reference object is dependent from the contextual frame (cf. Levinson, 2006). *Compositional* deixis are deixis, which are added in a spatial description, which is already defined and given (cf. Levinson, 2006). For instance, if one compares the utterances, *the nail at the edge of the table*, and, *the key is at this edge of the table*, the latter includes a more specific description of the spatial position of the nail in relation to the table and in the second utterance it is clear which edge of the table is meant. *Optional Origo* are deixis used in descriptions, in which the describing person acts as a centre of the coordinate system in space (cf. Levinson, 2006). An example of use of optional origo is the utterance, *the dog is at the left side of the box*, which is only true from the perspective of the describing person, which acts as the centre of the spatial situation (cf. Levinson, 2006). The first type of dexis, *central,* do not necessarily require a verbally pre-defined spatial relation, but it requires an additional dimension, mostly a visual configuration which can be supported by gestures. In contrast, *compositional* and *optional origo* can be based on a wholly verbal definition of the spatial relation, which can be understood without a visualisation of the space in which the objects described can be found. In the case of the use of deixis in different registers of language, central deixis seem to be more common in everyday language due to their requirement of a context. At first sight, both compositional and optional origo seem to exhibit characteristics of mathematics language i.e. requiring specific vocabulary (e.g. spatial prepositions), however, they are still context-oriented, but to a lesser extent than central deixis due to their higher degree of verbalisation. Therefore, spatial deixis can be regarded as elements which strenghten the aspect of how language and spatial-visual content are deeply interrelated and show the necessity of language to orientate oneself in space in a social environment. The issue of how spatial orientation can be represented in language is discussed in the upcoming section.

2.3.1.2 Spatial orientation and language

Although, at first sight, spatial orientation seems to be a wholly cognitive process, its concepts and operations can be given a social dimension by lexical and grammatical structures of languages (cf. Lehmann, 2013). When describing the localisation of the reference object, Levinson (1996) differentiates between three different frames of reference, which are also referred to as strategies for spatial orientation: *absolute, relative,* and *intrinsic* frames of reference.

In the absolute frame of reference, the individual chooses an invariable, context-free reference point, for instance, the four compass directions together with the directions "up" (sky) and "down" (ground) (cf. Levinson, 1996). An example of absolute frame of reference can be found in the phrase *the city hall is east of the church.* In this phrase, *east* is absolute and does not allow any interferences of the person's body position in the description of the spatial relation.

In the second frame of reference, *intrinsic*, an individual uses the structure of another object as a ground for spatial orientation: "in the intrinsic frame of reference the figure object is located with respect to what are often called *intrinsic* or *inherent* features of the ground object" (Levinson, 1996, p. 366). Consider the phrase *the central station is to the left of the cathedral*, were the structure of the ground *cathedral* is used to localise the position of the central station. If the ground does not have any structure at all, then this frame of reference cannot be applied in context. However, in this case the cathedral has a front, most probably where the main entrance door is located, and this serves for specifying the position of the central station. Whereas the intrinsic frame of reference describes the relation between two spatial objects, the next frame of reference can describe ternary relations.

In the *relative* frame of reference, individuals orientate themselves in space by choosing their body as the ground for orientation. In this case, the individual considers the structure of his body for orientation, such as the front and back side, the upper and lower side, and the right and left side. For example, consider the situation where a player is in front of a basketball stand and the ball is next to the stand. Let us assume that at first a player sees the ball to the left of a basketball stand. If the player turns 90 degrees anticlockwise around the stand from his original position, the ball would be behind the stand, thus the person is describing the spatial relation under consideration of their body position. As a matter of fact, this frame of reference can be considered as more flexible because the reference point can change its position or orientation in space.

Levinson (1996) states that whereas reference frames provide an adequate way to describe spatial orientation, they do not hinder ambiguity provided by the pragmatics of language. Consider the use of *behind* in the following phrases, *the car is behind the bus,* in which *behind* is semantically nonspecific concerning the reference frames. It could either mean that view of the car is obstructed by the bus (which could be equivalent to *the car is (driving) in front of the bus on the same lane and I am (driving) behind the bus on the same lane*) or the car is on the right or left-hand side of the rear of the bus from a passer-by's point of view. Therefore, when analysing spatial language, one should not only consider the conventional meaning of the elements carrying spatial meaning, but also the context in which these elements are used.

After a description of spatial language from a more linguistic point of view, the next section addresses the use of spatial language in mathematics learning and teaching. This is mainly done by focussing on current German mathematics curriculum specifications, since spatial language in German mathematics educational research has not been extensively researched.

2.3.1.3 Spatial language in German mathematics curriculum

Talking about space should be emphasised and supported in mathematics classroom, especially in geometry lessons, where students are required to describe objects and their location. In terms of curricular demands, the German mathematics curriculum provides just a few content-based specifications on space and figures with the emphasis on developing and learning spatial language in mathematics classroom. Such specifications in curricula address the need for students to be able to describe geometrical structures in the environment and to describe and reflect on characteristics and spatial relations between geometrical objects (e.g., position-relation; cf. Federal German Ministry of Education, 2004). Therefore, in general, the acquisition of spatial language is underestimated in German mathematics curriculum specifications.

Few German and international researchers (e.g., Wollring, 2012) have attempted to address the importance of spatial language in mathematics lessons. Wollring (2012) points out that vocabulary in the German spatial language is mostly borrowed and it has already got another meaning in everyday language. The phenomenon of borrowing in mathematics language from other varieties of language can also be intereprted as use of metaphors in the context of mathematics education (see Section 2.2.5). Consider, for instance, the term *Ecke* in the German language: *Ecke* in everyday language means corner, whereas in German spatial language, *Ecke* refers to a vertex of a solid in spatial geometry. A similar phenomenon can be observed in the English spatial language, where *edge* in everyday language can mean the border, e.g. *a path runs along the edge of the field.* This requires students to reconstruct and negotiate new meanings of known words in the new spatial context to learn spatial language successfully. Wollring (2012) states that merely the meaning of linguistic elements for expressing spatial orientation requires little, if any, negotiation in the students' spatial concept formation, for example, the distinction between left and right. The above facts show the need to raise awareness about the need for developing spatial language in mathematics classroom. The fact that many notions in spatial language are borrowed from everyday language increases the ambiguity and hence difficulties among students' understanding, especially if such new meaning are not explicitly negotiated in mathematics classrooms.

Overall, spatial language seems to play an important role in mathematics learning, especially in geometry lessons, in which students are required to describe objects and their locations, which are the functions of spatial language, described in Section 2.3.1.1. The amalgamation of spatial knowledge and linguistic elements in spatial language provides an appropriate approach for investigating the relationship between content and language learning in the present study. In this present work, spatial language denotes a variety of language which is not only limited to describing objects, but it should also include the language used to solve spatial tasks, to verbalise spatial thinking or strategies, and to orientate oneself in space in a social context.

2.3.2 Spatial task strategies and language

Following an overview of spatial language, a further attempt to provide another approach to the relationship between language and spatial abilities by focussing on language and the strategies for solving spatial tasks is provided in this section. As already described in Section 2.1.5, one of the most common differentiations in strategies for solving spatial tasks is *analytic* vs. *holistic* (cf. Barrat, 1953). In analytic strategies, learners focus more on specific characteristics of the object and use logical reasoning to solve the spatial task. In contrast to mental transformation and manipulation of objects in holistic strategies, spatial characteristics of objects and the underlying logical reasoning can be verbalised more easily (see Section 2.1.5.2). The stronger connection of analytic strategies to verbalisation and holistic strategies to visualisation can be observed in Schultz's (1991) categorisation of strategies for solving spatial tasks in *visual-holistic* and *verbal-analytic*. Hence, spatial language, whether in inner, external[16] or even in written form, seems to play an important role in analytic strategies of solving spatial tasks.

A similar argument is proposed by Schwank (2003) in her theory about thinking styles and the role of language. Schwank (2003) differentiates between two types

---

[16]   The terms *internal* and *external* are used synonymously with internal and external speech according to Vygostky (1993) (see Section 2.2.2).

of thinking styles: predicative and functional. The predicative thinking is more of a static modelling where the focus is on determining same or similar characteristics (for example, color or form) in invariant and structural connections. In contrast, functional thinking emphasises the transformation within thinking processes, thus focussing on the dynamic aspect of thinking. With regard to the role language in thinking processes, Schwank (2003) adds that language is important for solving tasks requiring predicative thinking processes. The recognition of characteristics and structural connections in predicative thinking processes in mathematics tend to be expressed using concepts which are easier to verbalise if learners can activate the corresponding knowledge adequately (cf. Schwank, 2003). In contrast to predicative thinking, learners solving a task using functional thinking processes are often unable to verbalise their procedural approach to the task adequately, not only due to the increasing cognitive demands of verbalisations, but also because language does not seem to be the main cognitive tool in accomplishing the respective task (cf. Schwank, 2003).

Schwank's (2003) approach to the two thinking styles is analogous to Barrat's (1953) distinction between analytic and holistic strategies for solving spatial tasks. The focus of predicative thinking on identifying features and structural relationships is typical for analytic strategies of spatial tasks. Similarly, holistic strategies are characterised by the dynamic process of transformation, which complies with the definition of functional thinking according to Schwank (2003). The emphasis on importance of language in both predicative thinking and analytic strategies makes the high similarity between Schwank's (2003) thinking styles and Barrat's (1953) spatial strategies plausible. However, Schwank's (2003) theoretical framework is more general in scope, since the objects of thought are not explicitly defined, as in the case of analytic and holistic strategies, in which the object of thought are spatial objects or their mental representation.

Language seems to play a more important role in certain strategies of solving spatial tasks over others. In particular, language used in analytic strategies to solve spatial tasks should exhibit certain characteristics, such as geometrical-spatial characteristics of spatial objects. Therefore, in addition to the overview of spatial language given in Section 2.3.1, the notion of spatial language should be extended by the language which is used for solving spatial tasks, negotiating spatial meaning in discourse and verbalising spatial thinking, whether it is of predicative/analytic or functional/holistic nature. Based on assumptions from previous literature (e.g., Schwank, 2003; Plath, 2014), students should find it easier to verbalise spatial thinking of an analytic nature, thus the verbalisation of holistic spatial thinking in spatial tasks is expected to be more challenging. Such facts strengthen the second approach to the relationship between language and spatial ability in this thesis, i.e. language is useful for solving spatial ability tasks in certain circumstances, based on the individual's thinking approach, whether analytic or holistic.

2.3.3 Summary: Interplay of language and spatial ability

In this section, I described the interplay of language and spatial ability based on theoretical findings from previous literature. Two approaches of linking language with spatial abilities were discussed. The first approach emphasised the use of spatial language in order to verbalise spatial thinking for understanding the underlying spatial conceptions. This focussed on the structure of spatial language and its underlying functions in spatial discourse: to describe spatial objects, their location in space or their location with regard to other spatial objects in space and to orientate oneself in space. The second approach was based on the potential use of language in developing strategies to solve spatial tasks. Previous literature shows that language could be influential in the development of analytic strategies for solving spatial tasks (cf. Section 2.3.2). It was agreed that the notion of spatial language should involve language used to solve spatial tasks, which can be realised either in oral or written form. Moreover, spatial language encompasses the aspect of verbalising spatial thinking or strategies, thus an analysis of spatial language provides a promising approach to investigate how students solve spatial tasks and their spatial knowledge. The different ways of how mathematical (also spatial) knowledge can be represented in mathematics education is discussed in the next section.

2.4 Representations of mathematical knowledge

In this section, different ways of representing spatial knowledge are discussed. In the first theoretical framework, the internal representations of mathematical conceptions, i.e. the nature of mathematical knowledge according to Sfard (1991), are introduced. This is followed by the second theoretical framework developed by Bruner (1966/1971), which focuses on the external modes of representations of mathematical knowledge.

2.4.1 Sfard's dual nature of mathematical conceptions

In her work about concept development in mathematics education, Sfard (1991) investigates the nature of mathematics conceptions, which are defined as "the whole cluster of internal representations and associations evoked by the [mathematical] concept – the concept's counterpart in the internal, subjective 'universe of human knowing'" (Sfard, 1991, p. 3). Sfard (1991) emphasises the need to focus on the cognitive processes from which mathematical concepts emerge and proposes two conceptions of mathematical concepts: *structural* and *operational* conceptions. Most mathematical concepts are known for their high degree of abstraction, whose mathematical definition, existence and properties are predefined by mathematicians and do not allow room for adjustments (cf. Sfard, 1991). The reference to the static nature of conceptions regarding mathematical notions is referred to as *structural*:

> Seeing a mathematical entity as an object means being capable of referring to it as if it was a real thing – a static structure, existing somewhere in space and time. It also means being able to recognize the idea 'at a glance' and to manipulate it as a whole, without going into details. (Sfard, 1991, p. 4).

In contrast, the other way of internalising mathematical concepts is by perceiving the mathematical concept as an array of actions in which the concept emerges. This means that instead of a static entity, an *operational* conception of a mathematical concept is of a dynamic nature and encompasses the operations and processes in developing the underlying concept:

> In contrast, interpreting a notion as a process implies regarding it as a potential rather than actual entity, which comes into existence upon request in a sequence of actions. Thus, (...) the operational [conception] is dynamic, sequential, and detailed. (Sfard, 1991, p. 4)

For a deeper understanding of the two conceptions of mathematical concepts, consider the concept of symmetry in geometry. Symmetry can be regarded as a geometrical characteristic of a form or object as in *a square is symmetric*. However, it can also be conceived as a process or set of actions as in the phrase *a shape is symmetric if it can be divided into two or more identical pieces*. In the former phrase of structural nature, the focus is on a static property of being symmetric or not. In the latter phrase, the focus is on the operational conception of symmetry, i.e. the process of dividing an object into identical parts, hence on the actions which should be performed by learners to develop or learn the concept of symmetry in an operational approach.

Sfards' (1991) theory evinces similarities with Schwank's (2003) thinking styles, which have been described in Section 2.3.2. The predicative and functional thinking style seems to conform with the structural and operational conceptions of mathematical concepts in Sfard's (1991) framework. However, whereas the focus in Schwank's (2003) theory is on thinking processes, Sfard (1991) elaborates more on the underlying mathematical concepts as results of the predicative and functional thinking style. Therefore, Sfard's (1991) theoretical work about the nature of mathematical concepts provides a suitable framework for this present study, in which students' spatial conceptions (as conceptions of spatial objects) in solving spatial tasks are investigated. In Section 2.3.2, I have indicated the plausible connection between the thinking styles from Schwank (2003) and Barrat's (1953) distinction between analytic and holistic strategies. Sfard's (1991) structural and operational conceptions can also be associated with analytic and holistic strategies respectively, whereby analytic and holistic strategies can be regarded as (mental) actions performed in spatial tasks which are of structural or static and operational or dynamic nature respectively. In the subsequent section, three representations regarding the external representation of mathematical concepts in mathematics education are discussed.

2.4.2 Bruner's modes of representation

In his research in developmental psychology, Bruner (1966/1971) proposed that mathematical knowledge can be represented in three modes of representation: the enactive, iconic, and symbolic. In the enactive mode, knowledge is represented by a set of actions which are required to achieve the underlying goal. The iconic mode is characterised by the focus on the pictures and graphics to represent knowledge. The symbolic representation is characterised by "a set of symbolic or logical propositions drawn from a symbolic system that is governed by rules or laws for forming and transforming prepositions" (Bruner, 1971, p. 45) and in contrast to other modes, it enables the representation of knowledge with a high level of abstraction. Regarding the use of the different representations, Bruner (1971) argues that there is no optimal sequence of the modes of representations during learning processes:

If it is true that the usual course of intellectual development moves from enactive through iconic to symbolic representation of the world, it is likely that an optimum sequence will progress in the same direction. Obviously, this is a conservative doctrine. For when the learner has a well-developed symbolic system, it may possible to by-pass the first two stages [the enactive and iconic]. (Bruner, 1971, p. 49)

Hence, rather than a sequence, Bruner's (1966/1971) modes of representation should be represented as interdependent entities, whereby knowledge can be represented differently in each mode and transferred from one to another. Bruner's (1966/1971) theoretical framework can be applied to describe the different representations of spatial knowledge. In the enactive representation, a set of real actions to represent spatial knowledge in solving spatial tasks can be the rotation of cubes or other spatial objects to visualise a particular rotation or movement. Students can also use a picture of the spatial object (e.g., supported by arrows) or a sequence of pictures (e.g., the spatial object in different positions) in order to visualise the rotation or movement of the spatial object in the iconic mode of representation. In the symbolic representation, spatial knowledge can be expressed in terms of mathematical symbols, words and phrases, such as representing the notion of rotation by using the degree(s) notation, by explicitly describing the involved actions of movement or by using rotation matrices. Regarding the representation based on the nature of spatial tasks, spatial tasks of a symbolic representation can include describing spatial objects and spatial relations and understanding their underlying spatial meaning based on words and phrases.

Most researchers (e.g., Bruner, 1971; Bruner et al., 1988; Plath 2014) agree that all three representations are important for development and learning of mathematical concepts. Figure 6 visualises a possible model for the representation modes in spatial tasks based on Bruner's (1966/1971) theoretical framework.

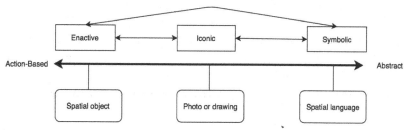

Figure 6   A visualisation of Bruner's (1966/1971) stages of representation under consideration of spatial knowledge

The double arrows between the three representations of knowledge, *symbolic*, *iconic* and *enactive* (cf. Bruner, 1971), in Figure 6 indicate the possibility of change in representation of knowledge. A symbolic representation of spatial knowledge can be represented by the highly abstract phenomenon of spatial language, whereas photos or drawings (for example, of spatial objects) can be used to represent spatial knowledge in iconic and enactive mode on a less abstract level respectively (see Figure 6).

2.4.3 Summary: Representations of mathematical knowledge

Two models about representing spatial knowledge were introduced in Section 2.4.1 and Section 2.4.2. First, Sfard's (1991) model about the internal representations in mathematics knowledge was introduced, wherein her theory was applied to the mental processes which can affect how spatial knowledge is represented in spatial tasks or spatial solving processes. However, the internal representation in Sfard's (1991) model is not enough to represent the whole complex phenomenon of spatial knowledge, since it does not reveal much about how and in which way spatial knowledge can be externalised. Therefore, Bruner's (1971) work about the external representations of knowledge – symbolic, iconic and enactive modes of representations – was considered as crucial for illustrating the different ways of externalising spatial knowledge. Hence, both models can be used for representing spatial knowledge both from an internal cognitive perspective and from an external social perspective, as it is illustrated in Figure 7.

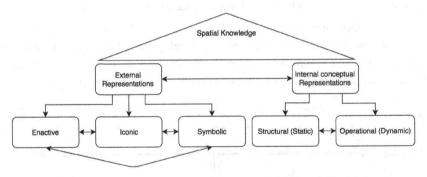

Figure 7   **Overview of the internal and external representations of spatial knowledge**

3. Methodology

This chapter presents the methodology applied in this present study which enabled to get insight into how students solve spatial tasks in spatial discourse. After having presented theoretical background of this present study in Chapter 2, the research questions are introduced in the first section of this Chapter, Section 3.1. From a methodological point of view, this research is characterised by "a methodology which enables it to describe how individuals interpret their actions and the situations in which they act" (Carr & Kemmis, 1986, p. 79). The underlying research paradigms adopted in this present study to approach the research questions is explained in the second section in this chapter, Section 3.2. In Section 3.3, a brief outline of the steps implemented during the research design process is illustrated for a deeper understanding of the subsequent chapter about design and implementation of the main study. In the last section of this chapter, Section 3.4, the research method is introduced and described, together with its underlying design and theoretical principles.

3.1 Research questions

Before proceeding with the methodological conceptions of this present study, I will recapitulate the aims of this present study, which were presented in the introductory section, Section 1.4, and present the underlying research questions for achieving results to the following main research questions of this present study:

How do different students solve spatial tasks? What role does language play in the description of spatial configurations in spatial tasks?

The first aim of this present study concerns the description of strategies which students employ when solving spatial tasks using language as a medium for externalising spatial knowledge. The second aim is to outline the obstacles which students encounter when solving these spatial tasks. These two research aims represent the qualitative core of this present study and are addressed in the following research questions:

(R1) Which strategies do students use to describe spatial configurations in spatial tasks?

(R2) Which obstacles do students encounter during the description of spatial configurations in spatial tasks?

Moreover, the qualitative data collected on strategies used by students and their spatial language in the spatial task solving process is analysed using quantitative methods. Frequency of students' use of the strategies identified in research question (R1) will be analysed regarding any dependency on background factors – students' language proficiency, spatial abilities, and sex. In the fourth goal, students' spatial language will be analysed using different structural approaches, which are used to investigate the structure of spatial language among students regarding any dependency on background factors of language proficiency, spatial abilities, and sex. Hence, further research questions for the two main research questions mentioned above are:

(R3) **Does the use of the identified strategies vary with the language proficiency, spatial abilties, and sex of the students?**

(R4) **To which extent can students' spatial language be analysed structurally and does its use vary with language proficiency, spatial abilities, and sex of the students?**

The last aim is to provide an adequate methodological approach for investigating the use of language in solving spatial tasks. Due to the fact that the research method applied in this present study has not been researched extensively in mathematics education research, a separate research question is formulated for this purpose:

(R5) **How can a suitable data collection method be designed for providing answers for the main research questions?**

In the framework of the first research question (R1), the strategies used by the students in the spatial task solving process are described, and, if applicable, compared to strategies from previous research. The second question (R2) focuses on the obstacles which students face during the solving of spatial tasks, especially the verbalisation of spatial configurations. The aim of the third research question (R3) is to investigate whether there is a substantial relationship between the use of the observed strategies and background factors of language proficiency, spatial abilities, and sex. The fourth research question (R4) sheds light on the nature of students' spatial language when solving spatial tasks by providing different structural approaches to analyse spatial language in discourse. The last research question (R5), which is primarily discussed in this chapter, addresses the methodological aspect of this thesis, which is the design of an appropriate method to collect data and provide interpretable results for the answering of the other research questions.

3.2 Research paradigm

The notion of research paradigm can be defined as "a model for collecting data and a theory for interpreting results" (Reeves Sanday, 1979, p. 527). From a methodological point of view, the research questions introduced in Section 3.1 indicate the integration of different methodological approaches, which characterises the research paradigm of the present study as a multiple paradigm. Whereas the development and the identification of strategies and obstacles in solving spatial tasks require data analysis with qualitative methods, the frequency of use of strategies and the structural analysis of spatial language among students under consideration of different factors require a quantitative data analysis approach. This situates the present study in a mixed methodology approach, since the multiple approaches are incorporated in the problem identification, data analysis, the transformation of collected data and their analyses to cater for the descriptive nature of research questions (R1) and (R2) and for the explorative nature of research questions (R3) and (R4). However, the methodological approach to answer the research questions situates the present study in qualitative research, which can be described as following:

Qualitative research is a situated activity that locates the observer in the world. It consists of a set of interpretative, material practices that make the world visible. These practices transform the world. They turn the world into a series of representations, including field notes, interviews, conversations, photographs, recordings, and memos to the self. At this level, qualitative research involves an interpretative naturalistic approach to the world. This means that qualitative researchers study things in their natural settings, attempting to make sense of, or to interpret, phenomena in terms of the meanings people bring to them. (Denzin & Lincoln, 2005, p. 3)

In the present study, language serves "as a kind of *window* to see indirectly what is happening in the student's mind" (Radford & Barwell, 2016, p. 276). The analysis of spatial language is an important tool to identify and investigate students' strategy development and obstacles in solving processes of spatial tasks in research questions (R1) and (R2):

(R1)   Which strategies do students use to describe spatial configurations in spatial tasks?

(R2)   Which obstacles do students encounter during describing spatial configurations in spatial tasks?

An analysis of student's spatial language requires an interpretative paradigm, in which the researcher puts himself or herself in the student's position in order to understand the student's way of thinking and being able to reconstruct knowledge and interpret these reconstruction processes to achieve the goals of the study (cf. Jungwirth, 2003). According to Krause (2016), the generated results from the reconstruction processes do not present "objective 'reality' (if existing) per se but specific aspects of it that always have to be seen as related to the specific setting and situation" (Krause, 2016, p. 55). Therefore, the researcher must design and reflect on a natural setting which is adequate for investigating student's thinking so that the researcher is able to reconstruct the student's performance and interpret it under consideration of the student's goals and intended meaning in the particular situation. This requires the researcher to assume that reality can be constructed in the same research process which is considered as an act characterised by constructivism (cf. Flick, von Kardorff & Steinke, 2004). From a methodological perspective, constructivism "assumes that human beings are knowing subjects, that human behavior is mainly purposive, and the present-day human organisms have a highly developed capacity for organizing knowledge" (Noddings, 1990, p. 7). Moreover, the descriptive research foci in the research questions (R1) and (R2) requires the researcher's methodologically controlled subjective perception to be an important component of evidence in the present study (cf. Flick et al., 2004).

Apart from describing strategies and obstacles in solving spatial tasks, the present study generates further theories about students' spatial thinking and language, which requires further attention in mathematics education (cf. Section 2.3). Therefore, research questions (R1) and (R2) build on the conceptions of grounded theory research (cf. Glaser & Strauss, 1967; Strauss, 1991). Grounded theory research is characterised by a set of characteristics, such as:

(…) constant comparative analysis, open and intermediate coding, theoretical sampling and saturation, theoretical integration of codes and categories, and memoing. Additionally, a crucial aspect of this research is the concurrent and continuous nature of data generation and analysis. (Teppo, 2015, p. 4)

In terms of a grounded theory research, the development and refining of categories for strategies and obstacles in students' spatial task solving process is a continuous process, which is characterised by the consideration of theory and by the generation of new theorical constructs to provide explanation for the observation and explanation of new phenomena. Another characteristic of grounded theory research is the consideration of theoretical sampling. Glaser and Strauss (1967) define theoretical sampling as the process of data collection to generate theory and then making decisions based on theoretical assumptions about the next steps for data collection, such as which persons shall participate in the study. Theoretical sampling is adequate for strengthening the rigour of the present study, especially for developing a theory-based empirical study which triggers a diversification of strategies and obstacles among students with mixed abilities. Theoretical sampling is also useful to investigate and generate further theory about the frequency of use of students' strategies and spatial language in relationship to other background factors possibly playing a role in the student's solving process, which are explicitly addressed in research questions (R3) and (R4):

(R3)   **Does the use of the identified strategies vary with the language proficiency, spatial abilities and sex of the students?**

(R4)   **To which extent can students' spatial language be analysed structurally and does its use vary with language proficiency, spatial abilities and sex of the students?**

Research questions (R3) and (R4) are based on an explanatory approach and require the development of a quantitative content analysis of the qualitatively collected data. Mayring (2015) describes content analysis as "a bundle of text analysis procedures integrating qualitative and quantitative steps of analysis (...)" (Mayring, 2015, p. 365). It is a research technique used for "the objective, systematic, and quantitative description of the manifest content of communication" (Berelson, 1952, p. 18). According to Guba and Lincoln (2005), there are two approaches to content analysis, which are based on different epistemological positions: *heurmeneutical* and *positivistic*. In heurmeneutics, often referred to as the constructivist paradigm, the emphasis is on understanding the meaning of human experiences in a socially constructed reality (cf. Cohen & Manion, 1994; Mertens, 2015). From a mathematics education perspective, the focus in the heurmeneutical approach is the understanding of language meaning in a social context, which in the case of the present study, should be more emphasised in the description of strategies and obstacles of students solving spatial tasks. The second position in content analysis arises from the paradigm of positivism, often referred to as the scientific method, which focuses on describing phenomena and highlighting predictions by means of observation and measurent (cf. O'Leary, 2004). In mathematics education research, the positivistic position entails the measurement, recording and quantification of aspects of language (cf. Mayring, 2015). Such aspects of language can be analysed using frequency analyses and other statistical techniques, which are considered as adequate to address the issues in the research questions (R3) and (R4).

In order to address the main research questions about how different students solve spatial tasks and about the nature of language in the underlying task solving processes, the present study is characterised by an integration of two different qualitative research approaches: interpretative and content analysis; triggered by the descriptive and explanatory nature of the underlying research questions. Since research questions (R1) and (R2) are the focus of this study, the main research paradigm of this present study is characterised as an interpretative paradigm. However, due to the explorative nature of research questions (R3) and (R4) the research paradigm of this study entails quantitative analysis of data which has been collected qualitavely.

3.3 Research design process

The research design process of the present study consisted of several steps which build up on each other. After having identified a research objective and problem, an adequate methodological approach for the data collection, which I refer to as *reconstruction method*, was chosen. The next step was to design the tasks for the reconstruction method, which were implemented in a pilot study. This was followed by an evaluation and data analysis of the pilot study regarding the research problem and also the data collection method. After refining the research problem, redesigning the tasks and the design of the data collection method, tests had to be developed for the sampling based on theoretical assumptions. The tests for the theoretical sampling were conducted in schools, and then evaluated to help with the selection of the tests persons in the main study. This was followed by the implementation of the main study, in which the chosen students were recorded whilst solving the task in the reconstruction method. The last step was the analysis of data collected during the main study. A visualisation of the steps during the research design process is observed in Figure 8.

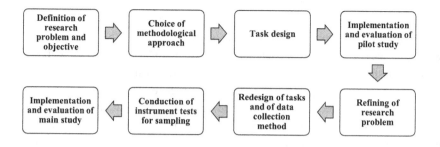

Figure 8   Steps of the research design process in the present study

3.4 Research method

As mentioned in the previous sections, one of the aims of the present study is to investigate spatial language when students solve spatial tasks. Two important foci of this present study are the description of strategies and of obstacles which the student face during the solving process. One of the main issues in designing this study was how to design appropriate spatial tasks in which both spatial language and spatial abilities play important roles for task solving, hence the last research question of methodological nature is addressed in this section:

(R5)   How can a suitable data collection method be designed for providing answers for the main research questions?

The underlying research method for achieving the above-mentioned goals should be deemed as appropriate for generating student's spatial language and its analysis to provide results for the present study's research questions. These factors led to the choice of a research method, which is referred to as *reconstruction method*. [17] In this section, the reconstruction method and its underlying design principles are introduced. Apart from the definition of the reconstruction method, this section includes a theoretical justification and a discussion about benefit and limitations of the use of this uncommon and under-researched research method in mathematics education to investigate the corresponding research goals.

---

[17]   This research method is referred to as *Rekonstruktionsversuch* or *Rekonstruktionsaufgabe* (cf. Wollring, 2012) or *Konstruktionsdiktat* (construction dictation) in German mathematics education.

3.4.1 Reconstruction method

The reconstruction method is a qualitative research method characterised by a particular communicative situation between students in a back-to-back spatial arrangement, who must solve an embedded task using language.[18] Tasks integrated in the reconstruction method are characterised by their dismantling in a series of steps and the preferred use of manipulatives in order to achieve the task's goal. The students in the reconstruction method are given different roles: *describer* and *builder*.[19] The describer is the one communicating the steps to solve the task and the builder is the one who has to act, implement and perform the steps after obtaining the information from the describer needed to solve the embedded task succesfully.

Figure 9   A visualisation of the reconstruction method and its embedded task

Due to the interweaving of the reconstruction method and the underlying task, and for a deeper understanding of this research method, a spatial task implemented in the reconstruction method is explained in more detail. Consider a setting in which two students are sitting in a back-to-back position in a classroom. One student, the *describer,* is given a figure or object, designed by the teacher or researcher, and he or she is instructed to describe to another student how this figure or object can be drawn or built up respectively. The listening student, the *builder,* must reconstruct, rebuild or reproduce the same figure or object using

---

[18]   Language can be realised in written or oral form. Since the focus of the present research is on spoken language, the latter form should be considered in the reconstruction method. A possible implementation of the reconstruction method using written language to communicate information is described in Wollring's (1998, 2012) work.

[19]   For a general description of the research method, preferably the number of participating students should be limited to two for easing the communication and the understanding of the underlying processes.

manipulatives according to the describer's description, hence he or she is assigned the role of a builder (see Figure 9). This task instruction is one way for implementing the reconstruction method and shows how deeply the task and the method are intertwined – the task for the realisation of the method and the method for the implementation of the task.

Only few studies by Wollring (1998, 2011) pointed out the the use of the reconstruction method in mathematics education research. The aim of the use of the reconstruction method in Wollring's work was to create a learning environment which supports learning of spatial and geometrical content. However, whereas a distinction between method and task is missing in Wollring's work, he implemented the reconstruction method for only two students at Grade 3, whereby the describer was instructed to draw a given object and the builder had to build the object as visualised on the desciber's drawing. Wollring (1998) used the data from this implementation of the reconstruction method to primarily focus on the difficulties the student encountered when drawing the three-dimensional object. Hence, this present thesis will thrive to give a more explicit and general definition of the reconstruction method whilst considering that it can be implemented using different tasks and rules, and to expand on potential of the reconstruction method as a qualitative research method for the investigation the role of spoken language in solving spatial tasks.

The observed data in the reconstruction method can be evaluated differently depending on the goal of the underlying study. Wollring (1998) differentiates between two types of evaluations: *result-controlling* vs. *communicative* evaluation. In the result-controlling evaluation, the main goal is to compare and control whether the describer's original object is identical with the builder's reconstructed object. This type of data evaluation emphasises more the comparison between the two objects at the end of the reconstruction method, hence *whether* the students manage to solve the task correctly or not (cf. Wollring, 1998). In contrast, in the communicative evaluation, the researcher focuses on the communicative processes taking place between the students during the solving of the tasks in the reconstruction method. The main questions focussed on in this type of evaluation deals with *how* the students manage to solve the task and *what* strategies they develop or obstacles they encounter during the task solving process (cf. Wollring, 1998). Therefore, the latter type of data evaluation is more adequate for the metholodogical approach to answer the research questions of this present study.

The focus on the communicative evaluation demands an interpretative approach (cf. Jungwirth, 2013), in which the researcher reconstructs the subject's knowledge constructed in interaction and interpret it accordingly. The reconstructed knowledge can be either declarative, in terms of spatial knowledge, or procedural, regarding how students approach the underlying spatial task. This explains the name of the reconstruction method, which should not be regarded as deriving only from the actual reconstruction of the model by the builder using manipulatives, but also by the reconstruction of knowledge based in the actions which reveal student thinking during the solving of the underlying task. On one hand, the builder is reconstructing knowledge mediated by language and translating it into actions, and on the other hand, the researcher is reconstructing knowledge about the communicative actions in interaction and cognitive processes which are supported by language.

For achieving the study's goals specified in Section 1.4, a more specific spatial task in the reconstruction method was developed to analyse strategies and obstacles in solving spatial tasks by focussing on the aspect of spatial language. The setting of the reconstruction method, which requires students to communicate with each other to share spatial knowledge, is ideal for investigating spatial language, which is the language used to verbalise spatial thinking during the solving of spatial tasks. The underlying design principles of the reconstruction method based on theoretical assumptions and to which extent the research method is adequate to reach the goals of the present study is discussed in the next section.

3.4.2 Design principles of the reconstruction method

The reconstruction method is based on different design principles which situate this research method in the qualitative research methodology. Due to the requirement of a specific task to describe the reconstruction method per se, the task instruction explained in the previous section (Section 3.4.1) serves as a potential task for the in-depth description of the underlying theoretical considerations in the reconstruction method. The design principles do not only provide a theoretical background for the reconstruction method, but also support its adequacy as a qualitative method to collect data when investigating the verbalisation and understanding of spatial thinking when solving spatial tasks, and to a broader extent the relationship between spatial ability and language. The design principles of the reconstruction method – communicative actions, manipulatives, verbalisation of thinking, and language use – are described in the next sections.

3.4.2.1 Communicative actions

Communication between students is an important characteristic in the reconstruction method. Communicative processes can be described as a collection of processes in which two or more students exchange, create and share meanings in shared contexts by using a language which both parties concerned can speak and understand. The reconstruction method creates a communicative situation in which productive as well as receptive communicative skills are essential and promoted. Productive skills require students to produce language, such as in speaking and writing. Speaking is the major skill playing an important role in the reconstruction method considered in this present study, because the spoken language is used to communicate knowledge needed to accomplish the task. There are several advantages of using oral communication in the reconstruction method. Spoken language allows immediate feedback, especially if the meaning of the utterances is not clear for the builder, and the negotiating of meaning of language or mathematical concepts. Moreover, oral instructions are more flexiable and adaptable in a new situation, such as the back-to-back students' seating in the reconstruction method. In contrast, receptive skills are those skills which do not require the student to produce language, but rather receive and understand the language, such as listening and reading. From the builder's perspective, listening skills are the most important skills for processing information and transforming it into actions for the construction of the object. From a methodological point of view, the communication in the reconstruction method should preferably not be limited by time for creating a more communication-friendly environment and therefore an effective student-student interaction. Moreoever, different students require different amounts of time for solving the spatial task, especially builders who need to understand and decode the utterances of the describer and most probably ask questions without any time pressure.

While designing the research method with its integrative task, it is important not only to create communication between the students, but also communicative actions in the reconstruction method. Communicative actions can be considered as a set of actions, which are coordinated via communication processes between two or more subjects to achieve a desired result. Such actions can be characterised as goal-oriented (cf. Aebli, 1980) which can be achieved by the cooperation of both subjects, and therefore both the communicative aspect and acting within the interaction are important principles for the design of the reconstruction method.

3.4.2.2 Hands-on manipulatives

Another design principle which is embedded in the task in the reconstruction method, is the use of *hands-on manipulative materials, manipulatives* for short, to support learning and communication of mathematics knowledge. Manipulatives denote the use of resources in mathematics classroom as medium to explore, explain, represent, interconnect, and exercise mathematical concepts and operations (cf. Lengnink, Meyer & Siebel, 2014). A manipulative is made up of a tangible material with which students can experiment and explore new relationships (cf. Lengnink et al., 2014). The use of manipulatives induces students to formulate questions or investigate issues to understand mathematical phenomena, such as calculations, processes and models, which can be used for argumentation and proofs (cf. Lengnink et al., 2014). The use of three-dimensional manipulatives in the reconstruction method is ideal for investigating spatial-geometrical issues, such as solids and shapes and can explain the prerequisite of the back-to-back position in this research method. The students must sit back-to-back because the manipulatives and the actions on these materials should not be seen by each other, otherwise the effectiveness of this data collection method, which in the case of this present study is based on the objective to create an adequate setting for students to verbalise spatial thinking and to understand spatial language, is not guaranteed. If only two-dimensional material is used in the reconstruction method, such as photos or drawings of three-dimensional objects, then the characteristic sitting position of the method is not necessarily required.

3.4.2.3 Verbalisation of thinking

One of the aspects of the relationship between language and thinking is the externalisation of inner speech in order to observe the students' way of thinking which is guaranteed by the design in the reconstruction method:

> This process of externalisation of inner speech can be facilitated by the design and implementation of an outer activity in which the child is encouraged to engage in social speech, i.e. speech addressed to other rather than oneself, which would enable the objective functional analysis of inner speech (cf. Vygotsky, 1993). (see Section 2.2.2, p. 41)

As Vygotsky (1993) states, an important criterion for the externalisation of inner speech is the design of an acitivity in which the student has the feeling that he or she is being understood and in which someone is listening to him or her (see

Section 2.2.2). The design of the reconstruction method fulfills this criterion, whereby the describer is externalising his or her thoughts by the medium of language addressed to another student, the builder, who is listening and acting in a concrete situation. Of course, the verbalisation of thinking and knowledge are cognitively demanding and influenced by other factors, for example, language proficiency. However, the reconstruction method creates a situation in which thoughts and knowledge relevant for the solving of the tasks has to be verbalised in order to be communicated to the builder. Therefore, the reconstruction method creates an authentic environment to verbalise one's thoughts, thus promoting the social dimension of thought. This research method does not only promote verbalisation of thinking, but also the structuring of thoughts and their selection when solving the spatial task, since not all thoughts are likely to be communicated to the other student. Moreover, it provides an adequate setting to investigate strategies students employ during the problem solving by focussing on the verbalisation of their thinking. These two characteristics show that the reconstruction method is an adequate qualitative research method to investigate the relationship between language and thinking to a broader extent.

3.4.2.4 Language use

The design of the reconstruction method requires describers to verbalise all the important information needed to accomplish the task successfully, i.e. language is the only way of representation for communicating knowledge in the reconstruction method. Considering integrated tasks which require students to describe spatial object and its construction allows the externalisation of students' spatial thinking to analyse spatial language, which is an important criterion of the design of this present study. Additionally, the reconstruction method allows not only the production of spatial language, but also its interpretation and understanding from the builders' perspective. As mentioned in the theoretical considerations about spatial language in Section 2.3.1, spatial language should be developed in a dialogue, and not in a monologue (cf. Coventry et al., 2009), which is fulfilled by the reconstruction method. Another characteristic which is induced by back-to-back position in the reconstruction method is the omission of non-predefined central deixis (see Section 2.3.1.1), unless the students specifically create a meaning for such words in a particular situation in their spatial discourse. The omission of central deixis enables the researcher to analyse students' spatial language, in which the spatial context needs to be constructed using language instead of using iconic or enactive actions to communicate spatial knowledge.

The above demands in the reconstruction method increase the complexity of the embedded task and demand higher language awareness from the students to describe objects more precisely and to reduce ambiguity. This might occur by using mathematical language, for example, geometrical concepts and spatial prepositions. However, the describing student must provide an adequate and understandable set of instructions to the addressee. In this way, students develop mathematics language awareness, thus using different registers and recognising the benefits of their use (see Section 2.2.4). The increasing language awareness in the reconstruction method should trigger students to appreciate and understand the complexity of communication and language and develop "a broader and deeper awareness of their own ongoing use of and relation to language itself (...)" (Hassanzadeh, Shayegh & Hoseini, p. 48 cited in Farias 2005, p. 48). Therefore, language in the reconstruction method is not only for communicative purposes and to pass on spatial knowledge, but also an object of reflection, whereby students and researchers are able to reflect on the used language and how the particular used linguistic elements can influence the processes or result.

The requirement of the reconstruction method to verbalise all the information needed to solve the task can be used as a diagnostic tool to assess student's ability to communicate knowledge. In particular, low language proficiency students can reach their linguistic limitation when describing in the reconstruction method. Linguistic limitations denote situations or moments in which students are unable to verbalise and communicate information required to solve the task successfully. The possibility of reaching the linguistic limitation is a more or less unique characteristic of the reconstruction method, which is created by the back-to-back position. This spatial arrangement results in redundancy of gestures (gestures cannot substitute words in the communication process of the reconstruction method) and omission of a common context, which are important for facilitating face-to-face communication among speakers. However, linguistic limitations should not only show and point out the students' linguistic barriers, but can also be used as an opportunity for further learning and increasing language awareness. When students reach their linguistic limitations, it might be useful to redesign or optimise the implementation of the reconstruction method by offering linguistic means to the students which can be used to overcome the linguistic barriers. This language support can be implemented as micro-scaffolding[20], which can be used to promote the development from everyday to mathematics language. An integration of microscaffolding in the reconstruction method enables the researcher to get insight of the following issues about students' conceptual and procedural knowledge in a central-dexis-free dialogue:

a) Which scaffolds or concepts do the students understand and can use correctly and a propriately in a particular situation?

b) Which scaffolds or concepts do the students perceive and adopt in their description? How do they react upon certain scaffolds in a particular situation?

The mirco-scaffolding integrated reconstruction method opens up new perspectives about language learning in this method. Due to the limited capacities of this study, the use of microscaffolding in the resonctruction method is not a research object of the main study in this thesis. However, the possible use of scaffolds in the reconstruction method supports the consideration that the reconstruction method can be used for language learning and awareness in mathematics classroom, and not only as a qualitative research method (as in the case of this present study).

3.4.2.5 Summary: Design principles of the reconstruction method

The design principles of the reconstruction method, communicative actions, use of manipulatives, verbalisation of thinking, and language use, illustrate the adequacy of this research method for investigating the relationship between language and spatial ability. The communicative actions demanded in the research method support the externalisation of thoughts, which are not only externalised in form of language, but also in form of actions on manipulatives. Hence, two external representations of spatial knowledge, the enactive and the symbolic representation, act as a medium for the analysis of student spatial thinking. The communication between the students supports the goal-oriented verbalisation of thoughts, since the builder is dependent on the describer's verbalisations of spatial thinking and knowledge. The use of manipulatives for the rebuilding of spatial objects (according to the describer's instructions) acts as an externalisation of the builder's interpretations of the spatial language developed by the describer. Hence, the research method allows the researcher not only to observe how stu-

---

[20]   The use of microscaffolding for implementing the reconstruction method was considered in the pilot study, which is discussed in Chapter 4.

dents, the describers, verbalise spatial thinking, but also how students interpret spatial language and translate it into actions on objects. For these reasons, the reconstruct method is deemed as adequate for investigating the aim of this present study, i.e. how students solve spatial tasks in which language plays a major role.

3.4.3 Limitations of the reconstruction method

The communicative and cognitive functions of language which are embedded in the verbalisation of thinking represent important design principles of the reconstruction method. However, the social transformation of the students' thoughts is generally not wholly, since subjects do not verbalise all of their thoughts. As pointed out by Vygotsky (1993), the inner speech is syntactically not as fully developed as the external speech, thus "forcing students to produce complete sentences or artificially joining fragmented utterances into grammatically complete 'communication units' may lead to misrepresentation of thought processes" (Charters, 2003, p. 76). This might be one of the limitations of this research method, similar to the qualitative method of thinking aloud (cf. Ericsson & Simon, 1980), since only part of the internal speech is externalised and hence this part serves as an object of analysis regarding the research objectives. However, it is assumed that the presence of another student, the builder, in the reconstruction method might reduce the effect of these constraints. In particular, the describer is expected to attempt to verbalise most of their thoughts concerning the spatial task, especially when considering that the builder has to perform actions based on the describer's instructions. One can argue that the describer's verbalisation in the spatial discourse is not purely his or her own strategical thinking, since it can be regarded as highly influenced by his language skills and also by the builder in a way that the describer focuses more on how to verbalise thinking to make it understandable for the builder. Although it is possible for the describing student to get influenced by the building students' language, observations in the pilot study, which is discussed in the upcoming chapter, Chapter 4, show that most often the other way round is the case, in which the describer's *dominance* in discourse affects the language used by the builder. However, in general whilst the back-to-back position in the reconstruction method creates new barriers, such as limiting communication and sharing of knowledge only via language, it also enables new opportunities for learning and developing strategies to overcome barriers, which are also a focus of this work. The next chapter provides a deeper insight into the implementation of the reconstruction method as a research method of the present study.

4. Design and Implementation

In this section, I provide the methodological framework explained in the previous chapter, by describing the design and implementation of this present study. The first section of this chapter, Section 4.1, is dedicated to the pilot study, whose results served as an important basis for developing and designing the main study, which is described in Section 4.2. Section 4.2 begins with an account of the reconstruction method applied in the main study (in Section 4.2.1), followed by a description of the task design (in Section 4.2.2), of the sampling and the instruments used for the sampling in the main study (in Section 4.2.3). This is followed by an account of the underlying quality criteria for qualitative research in Section 4.2.4. The implementation of the main study and data analysis for achieving the goals set by the research questions of this present study are illustrated in Sections 4.2.5 and 4.2.6 respectively.

4.1 Pilot study

Prior to the design and the implementation of the main study, a pilot study was conducted one year ahead of the main study in order to assess the feasibility of a full-scale study, test the adequacy of research instruments and further develop the research questions and research design of the main study. Although the research objectives and questions have been refined after the pilot study, the observations and the results from the pilot study have had a major impact on the design of and other decisions taken in the main study. The following sections give a brief outline of the design, implementation and results of the pilot study.

4.1.1 Aims of the pilot study

The major aim of the pilot study was to investigate language use for geometrical-spatial learning, whereby the emphasis was on the fifth-grade student's language use in geometry lessons. In particular, the pilot study investigated how students describe spatial objects, after having learnt the underlying geometrical concepts in geometry lessons. Other foci included an analysis of the linguistic means which students use when describing spatial objects, and how and which scaffolds students use in their descriptions. The adequacy, efficacy and the development of the research method, the reconstruction method, and of the integrated tasks were also a major objective of the pilot study, since the potential of the reconstruction method as a method of data collection has not been fully exploited in previous mathematics education research.

4.1.2 Design of the pilot study

In this section, the design of the pilot study, especially the design of the reconstruction method and its integrated task together with the manipulatives used in the pilot study, are discussed.

4.1.2.1 Research method of the pilot study

The reconstruction method was the method chosen to collect data in the pilot study for several reasons, especially under consideration of the assumption that it is adequate for the analysis of spatial language and development of strategies during the description of spatial objects. Moreover, the reconstruction method enables more authentic communicative actions, which support students to give a complete description, in contrast to a monologue or clinical interviews. Other reasons for the choice of this data collection method were discussed in Section 3.4.2. The design of the pilot study required more specifications about the implementation of the research method, since the reconstruction method can be implemented in different variations. Three different scenarios of the reconstruction method were developed, and required to be implemented in the pilot study for further development and the testing adequacy of the research method:

- Scenario 1 (restricting communication): The describer is the only student who is allowed to speak, whilst the builder only implements the instructed actions without conversing with the describer.

- Scenario 2 (control moment): Both students are allowed to converse freely with each other in a back-to-back position. At a certain point during the reconstruction method, the researcher shows the object which has been built so far by the builder to the describer, whereby the students remain seated in the back-to-back position. Once the describer perceives the reconstructed object, the researcher gives it back to the builder and the description can proceed. The aim of this intervention is to enable the describer to improve his or her description and detect errors or misinterpretations arising during the spatial discourse.

- Scenario 3 (language support): Both students are allowed to converse freely with each other in a back-to-back position. At certain point during the reconstruction method, the researcher offers the describer scaffolds in form of microscaffolding. The aim of this intervention is to provide linguistic support to the describer, if he or she reaches his or her linguistic limitation (see Section 3.4.2.4).

In contrast to Scenario 1, both Scenarios 2 and 3 included a two-way communication between the students in the reconstruction method. Of course, the above and other scenarios can be integrated together to form further scenarios, however, each of them entails distinctive features which might influence or support the describing process in the reconstruction method. All of the above scenarios in the reconstruction method with an embedded spatial task, which will be explained in the upcoming section, were tested in the pilot study.

4.1.2.2 Task design in the pilot study

As already mentioned in Section 3.4.1, the underlying spatial task and the reconstruction method should not be analysed separately, since they are deeply intertwined. While the reconstruction method is more of a set of rules for implementation, the task design consists of the researcher's instructions to the student and the needed manipulatives for solving the task. The following instructions for the spatial task in the reconstruction method was given by the research to the student(s) in the German language during the pilot study:

Researcher: *"Soon you [the describer] will get an object made up of these building cubes²¹, which can be linked together. You must give him/her [the builder] instructions how to build this object, so that he/she [the receiver] can reconstruct the same object. At the end, both objects must be identical".*

„*Gleich bekommst du [Beschreibender] von mir ein Objekt aus diesen Steckwürfeln, welche du [Beschreibender] zusammenstecken kannst. Du musst ihm / ihr [dem Nachbauenden] Anweisungen geben, wie er/sie das gleiche Objekt nachbauen kann. Am Ende, müssen beide Objekte identisch sein.*"

As the above task instructions indicate, the describer is given a spatial object made up of building cubes and he or she must describe how this object can be

---

²¹   For a visualisation of the building cubes used in the pilot study, see Table 3, which shows different objects made up from the mentioned building cubes.

rebuilt. The aim of the task, from the student's perspective, is that the builder reproduces the same object being described by the describer. The manipulatives consisted of building cubes of the dimensions 1.7 cm × 1.7 cm × 1.7 cm, which are available in different colors (see Table 3). One of the advantages of these building cubes is that students are able to plug them in in all directions. Moreover, these building cubes have a high stimulative nature and are used in German elementary schools to illustrate quantities, numbers, calculating and train spatial abilities in mathematics lessons. Further justifications for the choice of this manipulative will be discussed in the design of tasks in the main study (see Section 4.1.2.2).

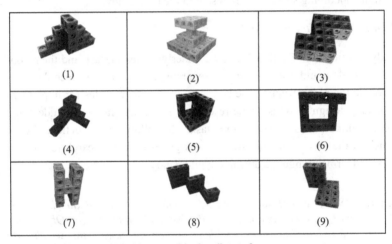

Table 3   Different spatial objects used in the pilot study

The objects used in the spatial task in the pilot study had different structures, in order to test the influence of structure of objects on the task solving process and improve the design of spatial objects for the main study. One of the criteria for designing the spatial objects was the adequacy of their structure for the students according to their mathematics performance. The spatial objects were varied in several properties, for example, variation in descriptive dimensionality of the spatial objects[22] (two-dimensional vs. three-dimensional description of spatial

---

[22]   Whereas all objects used are three-dimensional, the descriptive dimensionality should refer to the phenomenon of the students describing along the different dimensions. For instance, the one-cube depth of spatial object (6) makes the object two-dimensional in description, in contrast to spatial object (3), which requires the student to describe all its three dimensions.

objects) or concave vs. convex objects, hollow vs. solid, symmetrical vs. non-symmetrical, different spatial relations between parts of the objects (e.g., orthogonal or parallel) or two-dimensional characteristics (e.g., diagonal), congruent or similar parts of the object, similarity to structure of solids from everyday situations or solids known from geometry lessons (e.g., pyramid), and connected or non-connected parts of the object. Table 3 provides an overview of some of the spatial objects which were used for the task in in the pilot study.

4.1.3 Sampling and implementation

The pilot study was carried out with twelve students attending secondary schools in North-Rhine Westphalia in Germany: six students attending a Gymnasium and six students attending a Hauptschule[23]. Every student took part twice in the reconstruction method, once as a describer, and directly afterwards as a builder or vice versa. These students were picked by the researcher during the mathematics lessons and taken into a separate room where the needed apparatus and spatial setting was set up. The researcher gave the task instructions and explained the rules depending on the implemented scenario of the reconstruction method. Every scenario (1) – (3) (see Section 4.1.2.1) was implemented at least once with a different spatial object[24] (examples of spatial objects used are visualised in Table 3). At the end of the first reconstruction method, the students were instructed to change places and roles for the second round. At the end of each reconstruction method, the students were given time to comment or give feedback about the spatial task and the underlying spatial objects. All task solving processes were video recorded and transcribed for the evaluation of data and acquisition of results.

4.1.4 Results and consequences of pilot study

The evaluation of the pilot study focussed on the implementation of the reconstruction method, design of the spatial tasks and an intial overview of possible strategies used by students to solve the task. Before describing the observations of the evaluation of the pilot study, I would like to present an excerpt of a tran-

---

[23]   Both Gymnasium and Hauptschule are two types of secondary schools in Germany. However, on average students in Gymnasium are more high-achieving than students in Hauptschule. At the end of primary school, German students get a suggestion from teachers about which type of school (Hauptschule, Realschule and Gymnasium) they should attend according to their average performance in subjects.

[24]   The spatial objects in the reconstuction method were only given to the describer once both students were seated in a back-to-back position.

script from one task in the reconstruction method from the pilot study (in which the described spatial object is (5) from Table 3 in Section 4.1.2.2) to give an insight about the solving process of the task in the reconstruction method and the nature of spatial discourse which was considered for data analysis. In the following excerpt, the second column illustrates a translation of the children's utterances, which was recorded in German and are represented in the third column:

Speaker	English translation	Transcript in German
Describer:	(the describer counts the number of cubes) *Choose 33 stones.* (the describer starts counting again) *Choose 37 stones. Then make a surface... a four times four surface. Then make another one like it.* (the describer turns the original spatial object around so that the base area becomes the top surface). *Then link them as a crane house.*	(der Beschreibende zählt die Anzahl der Steckwürfel) *Such dir 33 Steine aus.* (der Beschreibende fängt wieder zu zählen an). *Such dir 37 Steine aus. Dann mach eine Fläche... eine vier mal vier Fläche. Dann mach dir noch so eine.* (der Beschreibende dreht den Körper so, dass die Grundfläche zur Deckfläche wird). *Dann verbinde sie so ein Kranhaus.*
Builder:	*A crane house? (...)*	*Ein Kranhaus? (...)*
Describer:	*Then it should look... like a half house! (...)*	*Dann musst so aussehen... wie ein halbes Haus! (...)*
Describer:	*The one from the quadratic angles... from this... Uhm I have found a mistake. Take a row first from one of them away... from that... so a quadrilateral and then you have to... one from the quadrilateral, but only from one you have to take away.*	*Das eine von den quadratischen Ecken... von diesem... Hhmm Ich habe einen Fehler entdeckt. Du musst erst bei dem einen musst du eine Reihe abnehmen... von dem... also ein Viereck und dann musst du von dem Viereck eine, aber nur von einem... musst du wegnehmen.*
Remark	(Intervention according to Scenario 3: Researcher shows the scaffold "perpendicular" (*senkrecht* in German) to the describer)	
Describer:	*Then you must, it... so the one, the long one... so this one where there is a four-triangle, there you have to... horizontal... it must be horizontal. And the other one perpendicular. Then you have to, at... at the end of the house... you must do a form and then link them.*	*Dann musst du die... also das eine, das lange... also dieses wo ein Vierdreieck ist, da musst du... waagerecht... es muss waagerecht sein. Und das andere senkrecht. Dann musst du an dem... Ende von dem Haus... musst du noch so eine Form machen und dann verbinden.*

An initial data analysis was carried out to find commonalities to create categories and codes which describe specific use of language leads to the establishment of strategies. In order to support a theoretical-based foundation of such strategies, the differentiation between holistic and analytic strategies was considered. However, the use of holistic and analytic strategies for identification of different strategies was not useful, because the analysed data was highly language-based and therefore more of an analytic nature from a first theoretical point of view. Consider, for instance, the above excerpt in this section, in which the describer's spatial language is the object of analysis. At the first sight, the student's utterances were based on describing the object's properties, hence a more analytic approach. However, this characterisation is not enough for differentiating between the different approaches undertaken by different students when describing spatial objects. A deeper analysis of the way how students think spatially in solving these tasks, which in turn is embedded in language, is needed in order to represent the wide spectrum of strategies beyond the simple differentiation of analytic and holistic. For instance, words, such as "crane house", "half house", and "quadrilateral" in the above transcript excerpt reveal which mental images the student activates to support their spatial thinking when solving the underlying spatial task. The description of identified strategies was further developed with reference to previous theoretical findings and is presented and described in detail in the results section.

The structure of the spatial object has had a strong influence on the linguistic means used by the students during the reconstruction method. Consider, for instance, the same transcript at the beginning of this sub-section, in which the describer uses the words "quadratic angles" (*quadratische Ecken* in German) (whereby he or she presumably meant quadrilaterals), "surfaces" and "four-triangle" (wherby he or she presumably meant four-angled or quadrilateral again, *Viererdreieck* in German). The use of these words might have been induced by the structure of the spatial object (5) in Table 3, and it is assumed that the same words would be less useful for objects having a different structure, such as objects (1) and (4) in Table 3. However, for a global comparison of the influence of the objects structure on students' spatial language the same objects are required to be used in a larger sample.

The implementation of the different variations of the reconstruction method was evaluated regarding the criteria of feasibility and adequacy regarding the goal of the study. An evaluation of reconstruction method with one-way communication, as in Scenario (1), showed that most of the builders could not follow the describer's instructions or decipher the meaning of words in context, because the communicative situation was less authentic. Hence, Scenario (1) hindered a continuous development of spatial language and the negiotation of meaning in the discourse. Moreover, hindering communication violates two important design principles of the reconstruction method, which are the embedded communication in task solving and increasing language awareness, which led to its rejection as a possible scenario for implementing the reconstruction method in the main study. In contrast, Scenario (2), in which two-way communication between the students and a control moment were allowed, proved to be adequate to promote the development of spatial language in communication, and to raise language awareness after the describer gets a glimpse of the reconstructed object for a short time during the reconstruction method respectively. An adequate timing for allowing the describer to get a glimpse of the reconstructed object during the research method is when the reconstructed object has a different structure than the original spatial object. This scenario has proven to be helpful in cases when the structure of the spatial object was relatively more complicated and hence increasingly cognitively demanding (e.g., spatial object (4) in Table 3). The control moment enabled the describer to restructure his or her spatial thinking and verbalisation (e.g., by change in linguistic means used in description, especially in case of communicating incorrect information or misinterpretation of information) and develop further strategies for overcoming obstacles which he or she potentially identified in the control moment.

Finally, the last scenario (3), characterised by the use of scaffolds has proven to be effective for examining content and language learning in spatial geometry. Scaffolds were used by the researcher when the describing students reached their linguistic capacity and were not always necessarily adopted by the describer. Since it shows how the reconstruction method can offer opportunities for language learning, it was concluded that this focus would go beyond the scope of the present research. Moreover, the use of micro-scaffolding interferes with an analysis of the student's actual spatial language for understanding students' spatial thinking. Therefore, the adaptation of scenario (2) in the reconstruction method (including a two-way communicative setting between the two students) should be reliable and adequate to reach the underlying goals of the main study of this present research.

As mentioned in the implementation of the pilot study, each student underwent the reconstruction method twice consecutively: first as a describer and then as a builder or vice versa. During data evaluation, a strong learning effect was noticed in the second round, where the describer in the second round used linguistic elements which were used by the previous describer in the first round. Moreoever, since the builder in the second round has had previous experience as a describer, he or she tended to influence the second describer's approach to the task, for example, by giving instructions about which aspects the second describer should describe. Therefore, in the main study, the exchange of roles among the same students in two consecutive spatial tasks in the reconstruction method was avoided, otherwise the describer's strategies would be highly influenced by the builder in the second round.

As already mentioned in the previous section, Section 4.1.3, the sampling group consisted of two groups of students from two different types of German secondary schools. An observation of student's use of language in the reconstruction method showed that the students' verbal skills differed substantially not only between the two groups, but also among the students within the same group. Although the mathematics teachers gave a vaguely judgement of the students' language skills from observations during mathematics lessons, a promising improvement for the main study was to assess the describer's language proficiency prior to the implementation of the study. The consideration of language proficiency as an influencing factor opened a discussion about which other factors can influence the spatial task solving process in the reconstruction method. This issue led to the establishment of two more factors, spatial abilities and sex, which are considered in the main study.

The design of the main study was based on a re-design and further development of the pilot study, which served as a basis for the investigation and testing of the research method and the design of the spatial task and identification of influencing factors in the solving process during the solving process. These improvements were adopted in the design of the main study, which is described in the upcoming section.

4.2 Design of main study

This section is dedicated to the description of the design of the main study of this present study. In the first section, the design about implementation of the reconstruction method under consideration of the possible different scenarios are described and justified. This is followed by the task design and its underlying demands. Next, the sampling of the study and the instruments used are described and the last part of this section consists of quality criteria and the implementation of the main study.

4.2.1 Reconstruction method in the main study

After the data evaluation of the pilot study concerning the scenarios of the reconstruction method (see Section 4.1.4), the implementing rules of the reconstruction method in the main study required further discussion. One decisive criterion was that such rules should not only make data collection feasible in the main study, but also help to increase the interaction between the students and to support the focus on the research aims and objectives. The first formulated rule is that the reconstruction method is implemented in form of a dialogue, meaning that the describer and the builder are allowed to converse and ask questions during the whole process without any constraints. This principle supports the idea that spatial language emerges stronger in a dialogue, because "the interaction in dialogue gives the opportunity for the students to participate more actively and creates a less artificial and less restricted setting" (see Section 2.3.1 as citied in Coventry et al., 2009). The same argument applies to the lack of designated time for solving the task in the reconstruction method. This means that students have potentially unlimited time to accomplish the task in the reconstruction method.

As already mentioned and justified in the previous section, the implementation of the reconstruction method in the main study should be based on facilitating communication between the students. Moreover, the data evaluation of Scenario (2) (see Section 3.4.3) in the pilot study was deemed as worthwhile for investigating the research aims of this present study, i.e. the evaluation of students' spatial language for investigating strategies developed and obstacles encountered during the spatial descriptions. In particular, the two-way conversation between the students allows an adequate setting for the development and analysis of spatial language (cf. Coventry et al., 2009). Moreover, the 'control moment' in Scenario (2) offers a productive intervention for the further development of the describer's spatial language or description and increased spatial awareness in the reconstruction method. Therefore, Scenario (2) was considered adequate for a possible implementation of the reconstruction method in the main study.

Another decision which had to be taken in the design of the main study is the number of times the reconstruction method is conducted with the same students.

As it has been described in the results of the pilot study (see Section 3.4.4), a strong influence in use of linguistic means between two students who exchanged roles in two consecutive task solving processes in the reconstruction method was noticed. To reduce this effect and promote the students' creativity and productivity in the main study, the students participated either only as a describer or only as a builder, but not both, in the reconstruction method implemented in the main study. Due to feasibility of the study and the inclusion of different structures of spatial objects in the embedded task, two tasks were implemented with the same students. In both task solving processes, the students are allowed to communicate with each other without time or any other restrictions, but only the reconstruction method in the first task should be implemented using the 'control moment' according to Scenario 2 (which allowed the structure of object in the first task to be structurally more cognitively demanding). Therefore, each student has participated twice (in the same assigned role) in the main study and is required to solve two spatial tasks which differentiate in terms of the underlying spatial object (see Table 4).

	Task A	Task B
Implementation of the reconstruction method	Two-way communication, control moment, and no specific time limit	Two-way communication and no specific time limit
Spatial object	Object A	Object B

Table 4   Overview of the tasks and reconstruction method implemented in the main study

4.2.2 Task design

As already indicated in Section 3.2.1, the spatial task embedded in the reconstruction method is characterised by the use of manipulatives, the design of the spatial objects employed in the task, and the researcher's instructions. Although the spatial task embedded in the reconstruction method is similar to the one in the pilot study, in this section, I will provide additional justification for characterising the task in the reconstruction method as a spatial task. In Sections 4.2.2.1 and 4.2.2.2, a description and justification of the choice of hands-manipulatives and the structure of the spatial objects used in the two spatial tasks in the reconstruction method are given. The task instructions and task demands are described in the Sections 4.2.2.3 and 4.2.2.4 respectively.

87

4.2.2.1 Choice of manipulatives

The choice of manipulatives to be used for the spatial task in the reconstruction method was influenced by several criteria. The manipulatives chosen had to be three dimensional, allow the construction of different structures of spatial objects and plugging-in between the component parts, and easy for the students to use in order to construct the spatial objects. Moreover, the chosen manipulative should support the development of the underlying spatial abilities in the task and the development of spatial language in discourse.

Figure 10   Manipulatives used in the tasks of the reconstruction method

The use of building cubes, also known as *snap cubes* (see Figure 10), for the building the original spatial object and the reconstructed spatial object in the reconstruction method fulfills the above criteria. Evidence of this is provided not only by the observations of the participating students using building cubes in the pilot study; results of previous research studies highlight the advantages of using building cubes for learning with manipulatives (e.g., Serbin & Connor, 1979; Fagot & Littman, 1979; Caldera et al., 1999; Casey et al., 2008). Caldera et al. (1999) show that there is a significant correlation between spatial abilities and construction using building cubes. In their study about observations of preschool children constructing objects using building cubes, Caldera et al. (1999) reported that preschoolers who showed interest in using building cubes performed better in spatial ability tests than other preschool children, who were not interested in playing with building cubes. Casey et al. (2008) also dealt with the correlation between spatial abilities and building cubes and concluded that building using cubes supports and improves the students' spatial abilities, especially the spatial components of mental rotation and spatial relations (see Section 2.1.3.3). However, building with cubes does not only support students' spatial abilities, but also the mathematical competencies in general. In their study about the relationship between manipulatives and mathematics achievement, Oostermeijer,

Boonen and Jolles (2014) noticed that students who played with building cubes in their free time achieved better results in solving mathematical text problems, rather than the ones who did not use building cubes at all.

The above findings show that building cubes are ideal as manipulatives for students solving spatial-verbal tasks in the reconstruction method, since it does not only serve as an adequate and easy-handling means for building and rebuilding of objects in the reconstruction method, but it also offers other advantages such as improving the students' spatial abilities and other mathematical competencies.

4.2.2.2 Design of spatial objects

An initial analysis of results from the pilot study in Section 4.1.4 showed that the structure of the objects plays a very important role in the type of language and spatial knowledge used in the reconstruction method to describe the object in the study. After selecting an appropriate manipulative, the next issue is the structure of the two spatial objects used in the main study.

In order to identify a wider spectrum of strategies or obstacles whilst students engage in the spatial task of the reconstruction method, the structure of both spatial objects should differ substantially. As mentioned in the design of the tasks in the pilot study, the design of the spatial object was characterised by several criteria (see Section 4.1.2.2). The use of different spatial objects enabled the researcher to explore the possible strategy choice of students during their spatial discourse and establish connections to the particular structure of the spatial object. For instance, the students tended to activate different everday or spatial knowledge influenced by the intended knowledge activation mediated by the object's structure. Based on such empirical observations from the pilot study, two spatial objects from the pilot study are used for the task design of the main study. As pointed out in the Section 4.2.1, each describer was required to solve two tasks in the reconstruction method in the main study. The spatial object employed in the first task is referred to as spatial object A, and in the second one as spatial object B. A visual representation of both objects, in form of a sketch and made up of manipulatives, can be seen in Figure 11 and Figure 12 accordingly.

Figure 11 Visualisations of spatial object A

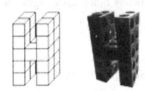

Figure 12 Visualisations of spatial object B

One of the common spatial property of the above spatial objects is their three dimensionality. Whilst all spatial objects constructed by building cubes are three dimensional, the notion *dimensionality* refers to the phenomenon whether the participants are required to describe only two dimensions (e.g., spatial object (6) from the pilot study in Table 3, see Section 3.4.2.2) or all three dimensions of the underlying spatial object. However, whereas spatial object A has a more complex frontal surface (in more than one dimension) in the spatial position represented in Figure 11, object B's front can easily be identified and more associated with the spatial-geometrical concepts of width, height and length. The perception of the multi-dimensional front of Object A is induced by the orthogonal spatial relation between two parts of the objects, whereby the orthogonality is constructed along the frontal and the horizontal axes. The ortogonality in spatial object B can be considered as two-dimensional, in the horizontal and vertical axis in space. Such assumptions require the students describing the objects to break it down in their description, which can be achieved in a number of different ways. In terms of break down possibilities, spatial object A seems to offer more different ways of break down than spatial object B. Moreoever, the structures of internal parts of the object can be compared to each other concerning spatial-geometrical properties, such as symmetry or preliminary notions of congruency or similarity concerning their dimensions. For instance, both spatial object A and B can be visualised as two identical internal parts depending on how the spatial objects are mentally broken down.

As mentioned earlier in the theoretical background about spatial abilities, mental rotation is considered to be an important ability to solve spatial tasks (see Section 2.3.3.1.1). Therefore, at least one of the spatial objects should be constructed in a way which requires the activation of mental rotation of the object for a solving the task successfully. This provides further justification for choosing the particular orthogonal spatial characteristic in object A, in which one part can be perceived as a rotation of the other one. This should lead to the use of (mental) rotation and its verbalisation in the task.

Furthermore, a preliminary analysis of spatial language in the pilot study has shown that students activate several metaphors whilst engaging in the spatial task of the reconstruction method. The activated mental images are influenced not only by the student's experience, but also by the structure of the spatial object in the task. Therefore, the structure of both spatial objects should induce the association with mental images, which the students are required to verbalise using spatial language. The important characteristics of the structure of both spatial objects, including the intended activation of knowledge and other properties, are summarised in Table 5.

Spatial object	Number of cubes	Description dimensionality	Intended activation of knowledge
A	20	3	Generation and activation of metaphors, break down of object in several parts, rotation of internal parts, spatial relation between internal parts, and spatial dimensionality.
B	22	3	

Table 5 Summary of the spatial objects' structure

4.2.2.3 Task instructions

Due to the consideration of two spatial tasks in the reconstruction method, which differ in their implementation, two task instructions and two spatial objects were required. The first part of the task instruction was the following[25]:

---

[25]   All task instructions have been provided in German. Hence both the German and English translations of the underlying task instructions are provided in the upcoming sections.

"In this experiment, you [the describer] will be given an object made up of these building cubes, which can be put together. You must give him/her [the builder] instructions on how to build this object, so that he/she [the builder] can reconstruct the same object. The colour of the building cubes is not important and whilst you [the describer] are describing you can also touch and move the object as you like, but the object structure must remain unchanged. At the end, the objects' structure must be identical".

"In diesem Experiment erhälst du [Beschreibender] ein Objekt aus diesen Steckwürfeln, die zusammengesteckt werden können. Du musst ihm / ihr [Nachbauenden] eine Anleitung geben, wie er / sie [Nachbauender] genau dieses Objekt nachbauen kann. Die Farbe der Steckwürfel spielt keine Rolle und während du [Beschreibender]beschreibst darfst du das Objekt anfassen und bewegen, aber die Struktur des Objektes muss unverändert bleiben. Am Ende müssen die Strukturen beider Objekte identisch sein."

As it can be observed above, the task instructions of the pilot study have been optimised and extended by further instructions. The participating students were instructed that they could perform real actions on the spatial object, which could support the observation of mental actions (cf. Pinkernell, 2003) and interpretation of language. In addition, the students were told that the colors of the building cubes of the reconstructed object did not matter, but the structure of both objects must be identical at the end. Due to the difference between the two spatial task implementation in the reconstruction method (see Table 4), further instructions were needed to distinguish between both implementations. The first task required the following additional instruction:

"You are allowed to talk to each other and ask questions. The only restriction is that you must not turn around. If he / she [the builder] is constructing a totally different object than the one you [the describer] are describing, at one point I will show you [the describer] what his or her [the builder's] object looks like, and you [the desciber] can take a glimpse of it and you can improve or change your description."

"Ihr dürft miteinander reden und Fragen stellen. Die einzige Einschränkung ist ihr dürft euch nicht umdrehen. Wenn er / sie [der Nachbauende] ein völlig anderes Objekt nachbaut, werde ich, zu einem bestimmten Zeitpunkt, dir [der Beschreibende] zeigen, wie sein / ihr [der Nachbauende] Objekt aussieht, und dann schaust du [Beschreibender] darauf und kannst deine Beschreibung verbessern oder ändern."

The second task in the reconstruction method required the following second part of the instruction:

"You are allowed to talk to each other and ask questions, but you must not turn around. Once you [the describer] think, that you are finished with the description, this experiment is over."

"Ihr dürft miteinander reden und Fragen stellen, aber ihr dürft euch nicht umdrehen. Wenn du denkst, dass du fertig bist, dann ist dieses Experiment vorbei."

In next section, the demands of spatial tasks introduced in this section and the underlying theoretical justifications are discussed.

4.2.2.4 Task demands

The task demands in the reconstruction method reflect the intertwining of language and spatial ability, both of which characterise the task solving process. An important demand of the spatial tasks in the reconstruction method is the externalisation of spatial thinking by verbalising spatial structures, relations, spatial positions and other spatial knowledge or actions. The describing students have to describe and communicate the spatial information required to build a particular geometrical object made up of building cubes by providing an instruction on how to build the spatial object. In return, the building student is required to decode the communicated spatial knowledge into actions in order to reach the goal of the task. Therefore, the solving of the task requires both students to have a strong command of (spatial) language and its underlying meaning and communication competencies.

In the first step, the describer must orientate herself or himself in the task and focus on the whole structure of the object by perceiving the object and identifying its spatial and geometrical characteristics. Due to the complexity of the object structure, the describer could mentally breakdown the object in order to provide a more adequate step by step description of the object. Under consideration of such information, the describer is required to choose an initial point to start describing the object or its internal parts and provide further information about how the building cubes or internal parts shall be assembled. The describer is required to identify two objects or parts of the object and describe their spatial relation accordingly, which activate cognitive processes such as the perception of figure and ground (see Section 2.1.2.1). The figure or the ground can consist of different items in the spatial situation of the reconstruction method. The figure or ground can be either a single building cube, or a part of the object consisting of more than one building cube, or the whole spatial object or the student's own body. Possible constellations of figure and ground in the task implemented in the reconstruction method are illustrated in Table 6. Other spatial-cognitive processes which the spatial task demands are the identification of spatial positions and spatial position of the objects in space (see Section 2.1.2.1).

Figure	Ground	Transcript[26]	Remark
Building cubes	Building cube	Describer: *"Do two cubes on each other"*	Both figure and ground consist of single building units.
Building cube	Object part	Describer: *"Okay, take a single block and plug it in the middle [of a part of the object]"*	The single block is the figure since it is moved with respect to the part of the object.
Object part	Object part	Describer: *"The three... three square... the three quadrilateral and the big one plug it [the three quadrilateral] on each other."*	The figure is the three quadrilateral which is one of the two object parts, and should be moved to be located on the other object part described as "the big one".
Object part	Human body	Describer: *"You have to... wait... do first one of these two twosome cubes stacked over each other."* Builder: *"Yes."* Describer: *"Then you have two lying in front of you."*	The figure is the twosome cube which is located or moved in a horizontal position with regards to the body position of the test person.

Table 6   Examples of constellations of figure and ground in the task of the reconstruction method

The demands mentioned above require describers to have a strong command of spatial abilities during the task solving process. Pinkernell's (2003) model of spatial abilities in mathematics education (see Section 2.1.3.4) provides the foundation for establishing further spatial demands of the task in the reconstruction method. The use of building cubes to construct spatial objects requires the describers to activate geometrical thinking and include geometrical characteris-

---

[26]   The transcripts in Table 6 originate from data collected during the pilot study.

tics of the object in their description, as in the second component of spatial abilities in Pinkernell's model (see Section 2.1.3.4). The describing students might reduce the spatial objects to their geometrical properties, such as width, depth, height, edges, faces in solid geometry, and length and breadth, shapes, symmetry in plane geometry (see Section 2.1.5.3).

Other components of spatial abilities required to solve the task include the mental and real construction of the object and the description of the object or parts of it and their transformation, such as movement or rotation, which falls within Pinkernell's (2003) first spatial component of spatial-visual operations. Whereas the describer tends to rather mentally construct the object by navigating in space from one building cube to another or one part of the object to another, the builder constructs the object using real actions on the building cubes after mentally processing the spatial information and actions received from the describer. Movement actions and transformations should also be operated on the spatial objects or their parts by both students, the describer rather mentally, whereas the builder first mentally and then through real transformation of the object or its parts. The movement of parts of the spatial configuration within the task, which are required to be verbalised rather than only visualised, is close to Thurstone's (1950) definition of visualization as an important component of spatial abilities (see Section 2.1.3.1).

Spatial orientation (according to Thurstone, 1950; see Section 2.1.3.1) is also required to solve the task in the reconstruction method. Both students have to orientate themselves in space under consideration of their body position or in relation to other objects in space. This is one of the frames of reference described by Levinson (1996), i.e. the relative frame of reference, in which the human body plays an important role in describing the spatial relation between a figure and ground and orientation of oneself in space (see Section 2.3.1.2). In particular, the space orientation of the builder is highly dependent on the spatial language of the describer. However, the reference to the human body as a means of spatial orientation does not need to be explicitly existent for the students to orientate themselves in space, as in Levinson's (1996) other two frames of reference, the absolute and the intrinsic frames of reference, which were described in Section 2.3.1.2. Table 7 shows examples of the three frames of reference which could be used in the solving of the spatial task in the reconstruction method.

Frame of reference	Transcript[27]	Remarks
Absolute	Describer: *"Firstly you build down so a tower to the top"*	In this utterance the describer uses the compass points together with "top" and "down" given by gravity. These directions are non variable in every spatial situation, which is characteristic for this frame of reference.
Relative	Describer: *"And whilst you are sitting, now go to the top and then right and build there three cubes there on the top."*	In this case, the speaker is the reference point and makes use of the structure of the physical object.
Intrinsic	Describer: *"Okay, take the four times four area and do two cubes on one corner and then plug in the three times three in a way that the two".*	Here, the describer uses two objects (*four time four area* and *two cubes*) and does not refer to the body for orientation, but rather to the three dimensional structure of the figure.

Table 7   Examples of frames of reference (according to Levinson, 1996; see Section 2.3.1.2) for spatial orientation in task of the reconstruction method

Therefore, the task demands both competencies from the sociolinguistic perspective and from the spatial ability perspective. The interplay between the visual (spatial object) and the verbal representations (spatial language) in the reconstruction method underlines the different forms of representation of spatial knowledge in spatial tasks (see Section 2.4.2).

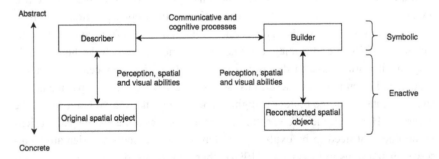

Figure 13 Communicative and cognitive processes demanded in the embedded task of the reconstruction method

---

[27]   These transcripts originate from data collected in the pilot study.

Figure 13 shows the processes which characterise the task in the reconstruction method. While the describer or the builder requires spatial and visual abilities to perceive the spatial object or perform actions on it, spatial language represents the externalisation of the underlying spatial knowledge in terms of communicative and cognitive processes of language (see Figure 13). Therefore, the continuous change of representation of spatial knowledge in the reconstruction method shows the importance of the processes of abstraction and concreteness in the spatial task. The nature of spatial knowledge which influences the process of description of the spatial objects is highly dependent on the structure of the object itself, which are described in the upcoming sections.

4.2.3 Sampling

In order to address the research questions, a number of different students had to be chosen for participating in the main study. Fifth-grade students (11-12 years old) were considered for sampling due to two reasons: (i) students at this age have just transitted from primary to secondary school, and therefore they are at a point which requires a relatively stronger development of language and mathematical skills, and (ii) students at this age are in a crucial phase in the development of geometrical thinking (cf. van Hiele, 1986; Piaget, 1970). Choosing students for sampling who are still developing their cognitive skills ensures less influence on students spatial thinking from other sources (e.g. influence from teacher thinking patterns) during their development of strategies, thus increasing the possible individuality within the paths of strategy development. Moreoever, the investigation of obstacles in solving spatial verbal-tasks is much interesting among students at this development phase, since it reveals what might hinder or affect the student cognitive development not only at this age but also in the next stages. The earlier the students' age in this development phase (12-15 years old in the last development phase of formal operations according to Piaget, 1970), the more distinctive obstacles of students solving spatial-verbal tasks are likely to be identified.

Additionally, an investigation of influencing factors in solving spatial tasks in the reconstruction method was required in order to construct the sampling for the main study. In the next sections, the sampling chosen for the data collection and for the implementation of the main study are introduced. After justifying the sampling by referring to theoretical foundations introduced earlier in this present work, the instruments for sampling are discussed.

4.2.3.1 Theoretical sampling

The sampling in the main study follows the idea of a theoretical sampling, which is based on the data collection process for the generation of theory (cf. Glaser, 1978; see Section 3.1). In this process, data was collected depending on the general research aim of the study and not based on a predeveloped theoretical framework from previous literature. After a first analysis of the collected data, the important key concepts and features were identified to refine the data collection processes and develop a theory within this process (cf. Glaser & Strauss, 1967). Such an analysis was conducted in the pilot study to observe the typical characteristics of the nature of students's solving of the task in the reconstruction method from a general point of view. Following the data collection, data transcription and analysis in the pilot study, important key features of the students' solving processes were identified and used to formulate observations and theoretise them. In the next step, the results from the pilot study were used for further refinement of the design of tasks, the validation of the conceptualised theory in the previous step and the focus on factors which can influence the use of student's strategies to solve the designed task in the recontruction method. According to Glaser and Strauß (1967), two important issues for the development of theoretical sampling are which groups of students are important for the data collection in the next step and for which theoretical purposes. The different language skills among students observed in the pilot study led to the emphasis on factors which are important for solving the designed tasks. Factors influencing the spatial task solving process, such as language proficiency, spatial abilities, and sex, were one of the major criteria for the selection of the sampling, which generated further research questions and refined the previous ones. According to the nature of task demands (see Section 3.5.2.3), the describers' spatial abilities and language proficiency are likely to influence the task solving process in the reconstruction method. Moroever, as a corpus of research about sex differences in spatial abilities has indicated, sex is another promising factor when considering solving processes of spatial tasks (see Section 2.1.6). The consideration of the above factors for sampling also provides *variability* in data collected about student's strategies, obstacles, use of spatial language and other concepts in the task solving process in the reconstruction method and allows "comparisons for similarities and differences, properties, and its relations to other concepts" (Teppo, 2015, p. 12).

The possible influencing factors – students' language proficiency, spatial abilities, and sex – led to the establishment of eight groups. These eight groups are characterised by the three dichotomies: high (L+) vs. low (L-) language proficiency, high (S+) vs. low (S-) spatial abilities, and male (M) vs. female (F) students (see Figure 14).

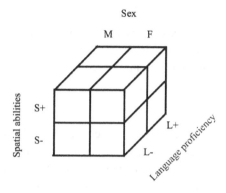

Figure 14 A visualisation of the sampling according to the factors of spatial abilities, language proficiency, and sex

For simplification purposes, the sampling was reduced to four groups, named A, B, C and D, each consisting of four students according to the different dichotomies. In group A, all students have high language proficiency (L+) and have high spatial abilities (S+). The second group of students, Group B, has high language proficiency (L+), but low spatial abilities (S-). Group C and D consist of students with low language proficiency (L-), but while students in the former group have high spatial abilities, students in the latter have low spatial abilities. To investigate the possible influence of sex on the task solving process, each group had to consist of male and female participants, hence, two males and two females in each group (see Table 8). Therefore, the total sampling consisted of sixteen student pairs (sixteen describers chosen according to sampling considerations illustrated in Table 8 and sixteen builders which achieved average scores in instrument tests), which should provide a wide range of possible constellations concerning the different factors in order to have a strong sampling to provide reliable results for the main research question.

	High language proficiency (L+)	Low language proficiency (L-)
High spatial abilities (S+)	Group A (M, M, F, F)	Group C (M, M, F, F)
Low spatial abilities (S-)	Group B (M, M, F, F)	Group D (M, M, F, F)

Table 8   Sampling under consideration of describers' spatial abilities, language proficiency, and sex.

The sampling for the main study required several instruments to be used in order to find suitable students which could participate in the implementation of this present study. These instruments for sampling are introduced in the upcoming section.

4.2.3.2 Instruments for sampling

The instrument tests introduced in this section were conducted prior to the implementation of the reconstruction method in order to determine the students' spatial abilities and language proficiency. The aim of the instruments for sampling was to build an adequate sample according to the dichotomies introduced in Section 4.2.3.1.

4.2.3.2.1 Instruments for language proficiency

In order to measure students' language proficiency, C-tests (cf. Klein-Barley & Raatz, 1984) were used as instruments for sampling in the main study. A C-test can be described as a collection of texts (mostly three to five texts) with gaps which must be filled in by students whose language proficiency is going to be assessed. During the design of C-tests the examiner must consider several aspects. Firstly, the topic of the tests should be authentic and predictable. In particular, the linguistic units of the tests should be appropriate for the students depending on their knowledge about the topic chosen for the C-test. One way to ensure that the syntax and the semantic of the text are adequate is to choose texts from school textbooks from the students' previous grade. For instance, if one wants to test language proficiency of fifth grade students, then one should use textbooks from the fourth grade to find an adequate text for the C-tests. Once this text has been chosen, it should be grammatically "damaged". One way of doing this is to leave the second part of each second or third word empty, starting from the second sentence of each text (see Figure 15). Figure 15 represents a translation of one paragraph from one C-test which was used in the main study to assess the students' language proficiency.

100

Der Schulbus

Jeden Morgen fährt Thomas mit dem Bus zur Schule. Die Kin__ stellen si__ in ein__ Reihe un__ warten, bi__ sie einste___ können. Thomas suc__ sich dan__ gleich ein__ Sitzplatz. Manch___ sind al__ Plätze beset__ und Thomas blei___ stehen. An ein__ Gurt kan__ er si__ festhalten. We__ der Busfah___ scharf brem___, stürzt e__ nicht z__ Boden. Vor all__ bei Regenwet___ ist e__ froh, da___ es ein___ Schulbus gi___.

The schoolbus

Everyday Thomas goes to school by bus. The chil___ line u___ and wa___ till th___ can bo____ the bu___. Thomas sear___ then f___ a se___. Someti___ all sea___ are occu___ and Thomas rema___ standing. Th___ he ca___ hold t___ a str___. When sudde___ the busdr___ brakes, h___ does n___ fall t___ the gro___. He i___ happy th___ there i___ a bu___, espec___ in rai___ weather.

Figure 15 An excerpt from a C-test designed for fifth grade students (above in German and below the translation in English)

The use of C-tests in major studies (e.g., in Prediger et al., 2013) for measuring student language skills shows that C-tests are useful for the objective measurement of language competencies. Moreover, C-tests provide an efficient and less time-consuming way to determine the language skills of a large number of students, which in the case of the present study, were whole classrooms. To assess the language competencies of the students, three German C-tests were designed specifically for fifth-grade students by the researcher, since there were hardly any generally accepted, standardised German C-tests to assess language proficiency of students at this grade. The texts to design the C-tests were chosen from fourth-grade German books to ensure that the tests did not entail difficult vocabulary or other complex linguistic aspects for the fifth-grade students. Further information about the designed German C-tests are provided in Table 9:

Test	No. of words	No. of items
CT-1	66	25
CT-2	67	27
CT-3	87	33

Table 9 Information about length of the C-tests and respective items

Due to time restrictions and other test factors, three C-tests with different themes were used to assess the language proficiency of the students in German. The tests were conducted in the presence of the mathematics teacher and the students were given the same stipulated time (6 minutes) to write each test consecutively.

4.2.3.2.2 Instruments for spatial abilities

Linn & Petersen's (1985) spatial ability model was chosen for developing the instruments to assess spatial abilities (see Section 2.1.3.2). This means that three different components or factors of spatial abilities were considered for its assessment: *spatial perception, mental rotation* and *spatial visualization* (cf. Linn & Petersen, 1985; see Section 2.1.3.2). This model has been considered for spatial ability assessment due to its establishment in recent research about spatial abilities (e.g., Büchter, 2011). Büchter (2011) states that Linn & Petersen's (1985) model offers a good description and clear-cut distinction of the mental processes and its transparency. Moreover, this model is ideal for the measurement of spatial abilities due to the availability of reference tests for each component: Water-Level-Task for spatial perception, Mental Rotation Test for mental rotation and Differential Aptitude Test – Spatial Relations Subtest for spatial visualization (see Section 2.1.3.2).

4.2.3.2.2.1 Water Level Tasks

The first reference test to be introduced, *Water Level Tasks* (WLT), was developed by Piaget and Inhelder (1958) and is useful to analyse the sub-ability of spatial perception according to Linn and Petersen (1985). Participants of these tests were given pictures which show different bottles in different positions and angles. The participants should assume that these bottles are closed and half filled with water and draw the water level of each bottle (see Figure 16).

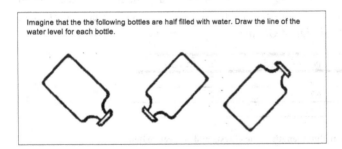

Figure 16 Task description and three test items in WLT reference test

To measure the spatial perception of students in the sampling of this present study, seven test items were chosen for the WLT test.

4.2.3.2.2.2 Mental Rotation Test

The mental rotation test (MRT) (cf. Vandenberg & Kuse, 1978; Peters et al., 1995) is the second reference test which was used to assess student spatial abilities. The MRT is based on the second factor of mental rotation in Linn and Petersen's (1985) model of spatial abilities. In this test, an initial three-dimensional object, referred to as the original object, is visualised from a particular perspective (on the left-hand side in the test item in Figure 17). The participants are provided with another four different pictures featuring three-dimensional objects from different perspectives. The student must identify which two of the four provided illustrations visualise the initial object from a different angle (see Figure 17).

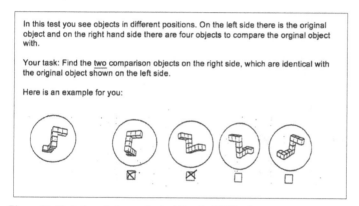

In this test you see objects in different positions. On the left side there is the original object and on the right hand side there are four objects to compare the orginal object with.

Your task: Find the <u>two</u> comparison objects on the right side, which are identical with the original object shown on the left side.

Here is an example for you:

Figure 17 Task description und a test item in MRT reference test

The MRT test in this study consisted of 10 items for which the students had 10 minutes to complete. Figure 17 shows the description of the task and one of the two items, which were given to students as an example at the beginning of the test. In the literature, there are other tests for measuring the factor of mental rotation in spatial abilities, such as *Bausteine Test* (BT) (cf. Birkel, Schein & Schumann, 2002). However, in these tests other factors of spatial abilities are also measured, such as the recognition of particular elements in given objects in BT Test. Due to the measurement of other spatial abilities in these tests, the MRT test was chosen over these tests, because it provides a more precise measurement of the factor mental rotation (cf. Büchter, 2011).

103

4.2.3.2.2.3 Differential Aptitude - Test-Subtest Spatial Relations (DAT)

The *Spatial Relation Subtest* (SR) from the *Differential Aptitude Test* (DAT) was developed by Bennett, Seashore & Wesman (1947), and is used to examine and measure students' ability to visualise movements and transformation of elements of objects. This reference test is used to measure the spatial ability component of spatial visualization according to Linn and Petersen (1985). In this test, a foldout of an object is presented to the students (see Figure 18) and four options of folded three-dimensional objects are visualised in the test. The students must mentally visualise the folding of the two-dimensional intial figure and choose the correct folded three-dimensional object from the four possibilities.

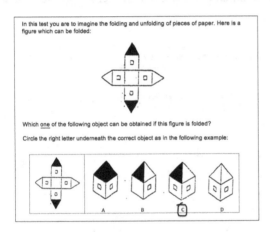

Figure 18 Task description and an item test in DAT:SR reference test

This reference test consisted of 10 items, which in total required approximately 10 minutes to be solved. Figure 18 shows the task description, together with a test item of the reference test (DAT: SR) which was given as an example at the beginning of the test.

4.2.3.2.3 Summary of instruments for sampling

The test instruments consisted of tests for assessment of language proficiency by using C-tests and reference tests to assess spatial abilities of students. The following tables, Table 10 and Table 11, summarise the most important information about the C-tests and reference tests respectively, such as the time required to solve the tasks in each test and number of test items in each test.

Test	No. of Items	Time (mins)
CT-1	25	6
CT-2	27	6
CT-3	33	6
Total	**85**	**18**

Table 10   Summary of the C-Tests to assess student's language proficiency

Test	Spatial factor	Time (mins)	No. of sample items	No. of items
WLT	Spatial perception	5	1	7
MRT	Mental rotation	10	2	10
DAT-SR	Spatial visualization	10	1	8
Total		**25**	**4**	**25**

Table 11   Summary of the reference tests to assess students' spatial abilities

Both spatial abilities and language proficiency tests were included in one booklet provided for each student, which consisted of a cover page, on which students had to write their name and class, followed by the three C-tests and the three spatial ability tests (see Table 12 for the structure of this booklet).

Test Name	Factor	Time (mins)
CT-1	Language proficiency	6
CT-2	Language proficiency	6
CT-3	Language proficiency	6
WLT	Spatial abilities	5
MRT	Spatial abilities	10
DAT-SR	Spatial abilities	10
Total		**43**

Table 12   Overview of the assessment booklet's structure

4.2.3.2.4 Scores in sampling tests

As already mentioned in Section 4.2.3, the student sample, which consisted of 16 describers, were chosen on the basis of three different dichotomies: spatial abilities, sex and language proficiency. Due to the requirement to fulfill the criteria established by the dichotomies, the instruments were conducted in different classrooms for choosing the sampling. After evaluating all tests of whole classrooms, students with above average or under average scores in both tests were noticed, and if their criteria fitted the requirements for sample group A (relative-

ly above average scores in reference tests and C-tests), B (relatively below average scores in reference tests and above average scores in C-tests), C (relatively
above average scores in reference tests and below average scores in C-tests) and
D (relatively below average scores in both tests), and in each group two males
and two females, then the students was chosen for sampling. The instruments
were conducted in different whole classrooms until all the sixteen students with
the requested scores and sex were found. The establishment of high or low
scores in tests was dependent on the average achievement of each participating
classroom in the respective tests, which varied from class to class, and was carried out by the average split. The scores in both tests of the students chosen for
sampling, including the respective average classroom scores, the standard devia-
tions, and z scores are illustrated in Table 13.

The average scores among all sixteen describers, who were all fifth graders but
from different classes or schools, are 38,44% and 67,88% in spatial ability tests
and in language proficiency tests respectively. These average scores are illustrated by the horizontal dotted lines in Figure 19 and Figure 20 respectively. The
corresponding results from the tests for spatial abilities and language proficiency
(see Table 13) are visualised in the following figures (Figure 19 and Figure 20)
according to the sample groups.

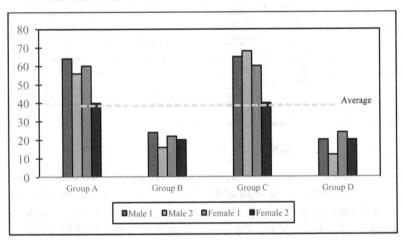

**Figure 19 Students' spatial abilities according to results of reference tests sorted in the sample
groups**

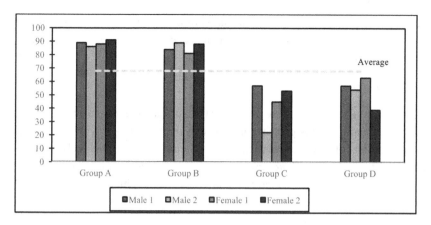

Figure 20 Students' language proficiency of test persons according to C-tests results sorted in the sample groups

Student	Sex	Group	Results of reference tests				Results of C-Tests			
			Score (%)	AVG[28]	STD[29]	Z[30]	Score (%)	AVG	STD	Z
D-1	M	A	64	30	18.81	1.81	89	60	28.63	1.01
D-2	M	A	56	36	13.36	1.5	86	76	14.67	0.68
D-3	F	A	60	31	16.37	1.77	88	65	14.49	1.59
D-4	F	A	44	30	18.81	0.74	91	60	28.63	1.08
D-5	M	B	24	31	16.37	-0.43	84	69	14.49	1.03
D-6	M	B	16	30	18.81	-0.74	89	60	28.63	1.01
D-7	F	B	22	48	20.01	-1.23	93	81	8.91	1.35
D-8	F	B	20	31	12.8	-0.86	88	69	20.57	0.92
D-9	M	C	65	48	20.01	0.85	57	81	8.91	-2.69
D-10	M	C	68	31	16.37	2.26	52	65	14.49	-0.9
D-11	F	C	40	31	16.37	0.56	51	65	14.49	-0.97
D-12	F	C	60	31	16.37	1.77	53	65	14.49	-0.83
D-13	M	D	20	31	12.8	-0.86	57	69	20.57	-0.58
D-14	M	D	12	30	18.81	-0.96	54	60	28.63	-0.21
D-15	F	D	20	36	13.36	-1.2	63	76	14.67	-0.87
D-16	F	D	24	31	12.8	-0.55	39	69	20.57	-1.46

Table 13 Student scores of factors spatial abilities and language proficiency in written tests.

[28] Average based on the overall achievement of the corresponding class.
[29] Standard Deviation based on the achievement of the corresponding class.
[30] Z-score or standard score.

Further information about the implementation of the test instruments in the class-rooms is provided in the Section 4.2.5, which is dedicated to the implementation of the main study.

4.2.4 Quality criteria

In this section, I will address the quality criteria of qualitative research, in order to judge the quality of this present study. The evaluation of quality of qualitative research consists of five criteria according to Steinke (2004): *inter-subject comprehensibility, indication of the research process, empirical foundation, explication of the limitations,* and *cohererency and relevance,* which are explained in the next sections.

4.2.4.1 Inter-subject comprehensibility

According to Steinke (2004), an appropriate inter-subject comprehensibility of a research process can be achieved in three steps. Firstly, the research process needs to be documented in order to allow third-persons to comprehend the pro-cesses which characterise the research process (cf. Steinke, 2004). This makes this quality criterion as one of the most important:

> The creation of intersubjective comprehensibility by means of a docu-mentation may therefore be regarded as the principal criterion or as a precondition for the testing of other criteria. (Steinke, 2004, p. 187)

However, documentation of the research process does not necessarily only refer to documentation of the research method itself, such as the information how the method has developed and in which context it is used, but it includes also the documentation of the researchers' role in influencing the research process:

> The necessity of documenting the researcher's *prior understanding,* his or her explicit and implicit expectations, results from the fact that these influence perception (for example, in observations), the choice or de-velopment of the methods used, and thereby the data collected and the understanding of the issue. (Steinke, 2004, p. 187)

After a documentation of the research method, the researcher is required to in-vestigate whether the research method has been carried out correctly, document transcript rules which had been adhered to during the data analysis process and a detailed description of the data analysis steps, which could, for instance, consist of the codified procedures which have been used in this step (cf. Steinke, 2004). The researcher must also document the information sources, such as the verbal statements of the interviewed participants and their intended meaning in the particular context, the researcher's own observations, hypotheses and interpretations (cf. Steinke, 2004). Steinke (2004) describes *interpretations in groups* as "a discursive way of producing inter-subjectivity and comprehensibility by dealing explicitly with data and their interpretation" (Steinke, 2004, p. 187), which are typical in hermeneunical approaches, such as in the case of this present study.

Due to a high volume of data which has been collected qualitatively, the student discourses in the reconstruction method were transcribed using simple transcription rules (mainly denoting questions [?], exclamatory statements [!], abrupt stops [X...], and long pauses in their discourse [...]). Being a study within an interpretative research paradigm, the collected data was analysed for possible multiple interpretations with other individuals to ensure comprehensibility for third parties and reduce subjectivity in data analysis. As already stated in previous sections, the pilot study played a major role in the design of the main study. A detailed documentation of the pilot study together with the implicit expectations, which developed throughout the whole research process, was provided in Section 4.1. Moreover, due to the lack of previous research about the research method used in this present study, a description of the reconstruction method, its underlying design principles and further observations regarding its implementation were documented (see Sections 3.2.1, 3.2.2 and 4.1 accordingly) for the reader to understand how and which data has been collected during the data collection process. Moreover, the outlining of the theoretical assumptions of the reconstruction method in Section 3.2.2 justifies the choice and development of this method used in the main study in order to answer the main research questions of this present research.

4.2.4.2 Indication of the research process

The criterion of indication deals with the appropriatness to the underlying research issue and it "is not only the appropriateness of the methods of data collection and evaluation but the whole research process that is being judged in respect to its appropriateness (indication)" (Steinke, 2004, p. 188). Regarding the appropriateness of the methods applied in this present research, the reconstruction method is appropriate to address the issue of the relationship between language and spatial abilities, in particular finding strategies and obstacles in which both language-based and spatial cognitive processes play an important role. Further justifications as well as decisions for not using pre-existing methods, such as the thinking aloud method, can be found in the discussion of theoretical foundations of the research method in Section 3.2.2. Moreover, an outline of possible limitations of the research method was provided in Section 3.2.3. Although the pre-structured students' back-to-back position in the research method creates new constraints, this aspect should not be classified as 'too severely constrained', because it complies with one of the aims of using the research method, which is the externalisation of spatial thinking. As a matter of fact, this phenomenon of the research method leads to the negligence of gestures for communication and other contextual information in spatial knowledge, which are relevant in the students' everyday context. However, one can argue that other assessment forms (e.g., written examinations) and situations in students' everyday context (e.g., phoning or writing text messages) do also emphasise more language use, rather than other dimensions of communicating knowledge. One of the main goals of this present study was the ultimate focus on the use of oral language for representing spatial knowledge of students, which is enabled by the appropriateness of the chosen research method. Therefore, the results should be deemed as appropriate in order to answer the research questions of this present study.

Another important aspect within this quality criterion of qualitative research is the indication of sampling strategy, which is induced by the question "To what extent is there an indication of such things as the cases and situations being investigated?" (Steinke, 2004, p. 188). The notion of theoretical sampling was an important design step in the research process of this present study. On investigating the strategies used among students describing spatial objects and the obstacles they encounter in this descriptive process, it was important to construct the sampling in a way to cater for different cases, for instance, not only high-performing students, but also low-performing students in language proficiency tests. Further details about the theoretical sampling and its justification can be found in Section 4.2.3.1.

Regarding the appropriateness of data evaluation, the qualitative data analysis consisted of cycles for establishing categories and coding the strategical moments in the describing students' discourses and the obstacles in the overall spatial discourse of both students in the reconstruction method. The justification for a greater focus on the describing students' utterances for coding strategies is initiated by the more active role in the reconstruction method which resulted in a higher contribution to the overall analysable spatial discourse. In contrast, the identification of obtacles required the evaluation of both describers' and builders' utterrances. Whilst evaluating the data for obstacles, the describer was regarded as the one triggering the obstacles, however, it was important to observe the builder's reaction or the result in order to ensure that concerning moment can be coded as an obstacle. Again, the established categories and the allocation of data to categories in the evaluation process process were tested for appropriateness in interpretation groups as indicated in Section 4.2.4.1.

4.2.4.3 Empirical foundation

Empirical foundation is another important criterion for evaluating the quality of qualitative research. This criterion emphasises the importance of the data in order to create new theories or test hypothesis in qualitative research (cf. Steinke, 2004):

> *Theory-formation* should happen in such a way that there is a possibility of making new discoveries, and questioning or modifying the investigator's prior theoretical assumptions. Theories should be developed close to the data (for example, the informants' subjective views and modes of action) and on the basis of a systematic data analysis. (Steinke, 2004, p. 189).

Steinke (2004) emphasises that the verification or falsification of hypotheses for testing of theoretical assumptions should also be based on empirical data. One way of providing empirical foundation for research is the use of objective hermeneutics or grounded theory (cf. Steinke, 2004). Both phenomena play an important role in the empirical foundation of this present study, which considers a hermeneutical approach for understanding the meaning of language used by students in solving spatial tasks and the grounded theory for developing new theories based on existing ones from previous frameworks about students' strategies or obstacles in solving spatial tasks.

4.2.4.4 Limitation

Limitation is a criterion which addresses the limits of research process regarding its application, generalisation of its results and the developed theory (cf. Steinke, 2004). In order to investigate the possible limitations of the results, "an analysis of the further conditions (contexts, cases, investigated groups, phenomena, situations and so on)" (Steinke, 2004, p. 189) is required. In this analysis, the researcher is required to prove which conditions must always be fulfilled in order to being able to generalise the obtained results:

> There must also be a clarification of what conditions must be fulfilled, as a minimum, for the phenomenon described in the theory to occur. At the same time, aspects that are incidental, and – from the point of view of theory – irrelevant, are filtered out. This can be discovered through the introduction, omission and varying of conditions, contexts, phenomena, and so on that are relevant to the creating or influencing of the research issue. (Steinke, 2004, p. 189)

Regarding the goals of the present research, the consideration of different sampling groups according to the three possibly influencing factors (see Section 4.2.3.1) contributes to the reliability of the developed theory concerning the identification of strategies and obstacles in the task solving process. In the task design of the present study, the choice of two spatial objects from an unlimited number of structures could point out a limitation of this study. Whilst the use of several different spatial objects in the pilot study helped to justify the choice of both spatial objects in the main study (see Section 4.2.2.4), one has to consider the feasibility of the study in relation to the sampling groups in the theoretical sampling. Moreover, the choice of manipulatives in task design plays an important role in the student's use of spatial language (see Section 4.2.2.2). Consideration of other manipulatives to construct the spatial objects in the spatial tasks in the reconstruction method, such as cubes which cannot be plugged in together, or the use of triangles as building units, would not fulfill the criteria for an adequate spatial task (e.g., the use of triangles would limit the generation of different structures of spatial objects extensively).

4.2.4.5 Coherency and relevance

Coherency and relevance are two other critera which are important for the evaluation of qualitative research. Coherency deals with the understanding of the theory which has been generated in the research process, including the processing of possible contradictions in data or interpretations (cf. Steinke, 2004). Relevance concerns the pragmatic usefulness of the generated theory in qualitative research, which can be induced by the following questions: "Is the research question relevant?" (Steinke, 2004, p. 190), "What contribution is made by the theory developed?" (Steinke, 2004, p. 190), and "Does the theory facilitate the solution of problems?" (Steinke, 2004, p. 190). Such questions are important for the researcher to locate his research and its underlying results in the mathematics educational research and reflect upon the consequences of such results in the research of this discipline. The issue of language and content learning was described as highly relevant in research of German and international mathematics education – especially of language and spatial learning, which has not been researched adequately in German mathematics education research up to now. The theory generated in this present research helps to address the complex externalisation of spatial thinking via language from a students' perspective, whereby students' strategies and obstacles in solving spatial tasks using language are identified. Whereas it does not offer direct solutions, the generation of theory helps to understand better the complexity of the phenomena investigated in this present study. An identification, description and understanding of problems in this present study serves as prequisite for development of solutions and support in the underlying research domain of mathematics education.

4.2.5 Implementation of main study

The instrument tests were conducted in five different fifth-grade classrooms of three different secondary schools – one Realschule and two Hauptschulen – in the state of North-Rhein-Westphalia in Germany. All students attending these classes had to sit for the instrument tests which were supervised either by the researcher or a teacher or both. Even though a description of the task and an example were given at the beginning of each test, the teacher explained what to do in each task to ensure that the students understood the task. The tests were written in a double lesson to ensure there was enough time for the introduction and any possible questions or explanations.

Each test booklet consisted of a cover page, three subtests for language, and three subtests for spatial abilities (see Table 12). After filling in the personal information on the cover page, the students were required to fill in the gaps of all three C-tests after each other. Once the students have finished the C-tests, they had to wait for the teacher's instructions to start with the first spatial ability test. Between each spatial ability test the students had to wait till everyone in class had finished, because an explanation of each spatial ability test was provided by the teacher. At the end, all booklets were collected to continue with the data evaluation and analysis to choose the students for the reconstruction method. The

113

collected booklets were scored according to a specific system. Each test item in the booklet was assigned a point, if it was answered correctly. Hence in the C-tests, a correct answer in the gap was awarded 1 point, otherwise 0 points. The answers in the gaps had to be also syntactically correct, otherwise these were classified as incorrect.

Name	Test 1	Test 2	Test 3	Total	%
	25	27	33	85	100%
	8	15	15	38	44,71
	19	15	15	49	57,65
	13	7	15	35	41,18
	17	11	15	43	50,59
	12	21	15	48	56,47
	12	14	8	34	40
	19	19	0	38	44,71
	21	16	6	43	50,59
	21	19	4	44	51,76
	23	21	0	44	51,76
	21	22	2	45	52,94
	24	22	0	46	54,11
	18	10	21	49	57,65
	24	20	9	53	62,35
	21	19	18	58	68,23
	22	21	18	61	71,76
	19	19	24	62	72,94
	20	21	24	65	76,47
	22	24	19	65	76,47
	21	23	23	67	78,82
	23	22	24	69	81,18
	23	25	22	70	82,35
D-3	23	24	28	75	88,23

Table 14: C-test scores of one specific class for sampling

Name	WLT	MRT	DAT	Total	%
	7	10	8	25	100
	1	0	1	2	8
	1	0	2	3	12
	1	0	2	3	12
	1	1	2	4	16
	2	2	0	4	16
	2	1	2	5	20
	1	0	4	5	20
	4	0	2	6	24
	1	1	4	6	24
	3	1	2	6	24
	1	5	0	6	24
	5	0	2	7	28
	3	2	2	7	28
	2	2	3	7	28
	3	4	2	9	36
	4	4	1	9	36
	2	4	3	9	36
	3	6	1	10	40
	5	4	1	10	40
	4	6	3	13	52
	7	8	0	15	60
D-3	6	6	3	15	60
	4	9	4	17	68

Table 15: Reference test scores of one specific class for sampling

In the case of the spatial ability tests, two of which were multiple choice tests, there was also a point for each test item, if the corresponding task was answered correctly. In the case of the MRT test, in which the students must choose two perspectives of the rotated object, both answers had to be correct in order to achieve a point per test item. In the WLT test, a test item was considered as correct if the line was within five degrees of true horizontal. After the booklet correction, the results from the three C-tests of each student were added and sorted according to the total points. The same procedure was applied to the results of the spatial tests. All the results from the the tests were gathered in Excel and provided an overview of the results of the classroom achievement. Moreover, descriptive statistics (average split and standard deviation) for both C-tests scores and the reference tests scores were used for the organisation of the results. Table 14 and Table 15 show the results of the tests in one specific class, from which a describer, for instance, D-3 was chosen as a participant for the reconstruction method in the main study. The results, as the ones illustrated in Table 14 and Table 15, were used to determine which students in the particular classroom has scored over average or below average in C-tests and in spatial ability tests, in order to choose the sampling illustrated in Section 4.2.3.1. For instance, D-3, a female student who scored above average marks in the C-tests and had an over average score in the spatial ability tests would fit in the category (L+, S+, F) and was therefore considered as a potential candidate for the reconstruction method in the main study. The selection of the candidates was based on the theoretical sampling and did not exceed eight participants in each dichotomy.

Since the students attended different schools, the same procedures (tests implementation, evaluation and choice of students according to theoretical sampling) were repeated for every classroom, from which students were chosen as describers in the reconstruction method in the main study depending on the scores in sampling instruments. Each of these students needed to be accompanied by another student which would act as a builder in the reconstruction method, needed to construct the described object. These builders were chosen in a way that they mostly achieved average results in both tests in order to reduce any difficulties which might emerge in the reconstruction method due to different achievement in spatial abilities or language proficiency. The students which fitted the sampling requirements described in Section 4.2.3.1 were then invited to take part in the reconstruction method.

The spatial task solving processes were recorded either at school or at university, where some student groups have been visiting to take part in the mathematics lab "math-checkers" at the University of Duisburg-Essen. Both actions of the describer and the builder were recorded on video. In each recording two cameras were used, one for capturing the describer perspective and one for the builder in the reconstruction method. Following the preparation for the necessary spatial arrangement of the participating students, the researcher gave a brief introduction of the aim of the study and explained the rules of the reconstruction method and the first task instructions were given. The researcher had to explain the roles to each student and advised not to turn around if not instructed by the researcher to do so. Moreover, the builders were provided with enough building cubes in order to be able to build the described object. A trial of the task in the reconstruction method was performed using a simple spatial object made up of seven cubes as a specimen to ensure that the students understood the tasks properly before starting with the actual task of the main study.

Once the students were seated in a back-to-back position, the researcher signalised the beginning of the task in the reconstruction method, gave the original figure to the describing student. At this point the describing students were instructed to start with his or her description of the spatial object. The describers, which were chosen according to the sampling, had to describe two different objects, object A and object B, previously designed by the researcher (see Section 4.2.2.2). This means that the two tasks were solved in the reconstruction method by the same students, but the rules of the reconstruction method differed according to different scenarios (see Table 4 in Section 4.2.1). At the end of each task, the participants were allowed to turn around and control whether the original and rebuilt object were identical.

4.2.6 Data analysis

The data collected during the implementation of the main study needed to be analysed using multiple analyses – inductive and deductive analyses – depending on the underlying research question. The data analysis process of this present research can be described as "working with data, organizing it, breaking it into manageable units, synthesizing it, searching for patterns, discovering what is important and what is to be learned, and deciding what you will tell others" (Bogdan & Biklen, 1982, p. 145). In order to work with data, a transcription and organisation of the data was required, followed by its breakdown and synthesis into uterrances, phrases or words in spatial discourse depending on the nature of data analysis. A detailed description of the data analysis steps, including the organisation of data, inductive and deductive data analysis, and presentation of the results is described in the following sub-sections.

4.2.6.1 Transcription and organisation of data

The first phase of the data analysis process consisted of the transcription of data collected during the solving of the spatial tasks in the reconstruction method. Both the utterances of the describers, builders and the researcher in the recorded videos were transcribed using simple transcription rules[31]. The sixteen participating students were anonymised using letters D (for Describer) or B (for Builder) according to their assigned role followed by a number indicating the student pair, for example, D-1 or Describer 1 stands for the first describer in the reconstruction method and B-1 or Builder 1 for the first builder in the reconstruction method. If the meanings of utterances in the transcripts were ambiguous, students' gestures provided an alternative way to reconstruct the actual intended meanings.[32] An additional representation of knowledge which was useful for understanding the actions in the reconstruction method is the visualisation or a set of visualisations of the reconstructed object. However, due to the high volume of data, the gestures and other aspects, such as intonation of students' utterrances, and visualisations of the reconstructed object per turn during the reconstruction method were not transcribed. Visualisations of the builders' reconstructed objects were provided at the end of each reconstruction method transcript (see Appendix B) and during the exchange of the objects in reconstruction method A (see Section 4.2.1). The reconstruction method transcripts served as a medium for analysing the data in the next steps. The data analysis was carried out in German, but the results of the analysis are represented in English. Therefore, students' utterances from transcripts are provided both in the English and in the German language in the results section of this thesis. The first data analysis, inductive analysis, is described in the next sub-section.

4.2.6.2 Inductive analysis

An inductive analysis can be described as an approach which "primarily use[s] detailed readings of raw data to derive concepts, themes, or a model through interpretations made from the raw data by an evaluator or researcher" (Thomas, 2006, p. 238). From a grounded theory approach, this process of data analysis

---

[31]   The considered transcript rules were the following: [?] for questions with the corresponding intonation, [!] for exclamatory statements, [(...)] for continuation of episode, and [X...] for showing abrupt stops (if the word X is not complete) or [...] for long pauses.

[32]   In general, the task demands tended to reduce the ambiguity of the utterances, since the use of central-dexis is very limited in the reconstruction method.

enables the generation of new theory from the collected data (cf. Strauss & Corbin, 1998). Thomas (2006) identified three main goals of inductive analysis approaches based on data transcripts, which are applicable in grounded theory research:

1) Summarising the extensive and varied raw text data into a brief format.
2) Establishing links between the research goals and findings from the raw data and ensuring that these are transparent and justifiable.
3) Developing a model or theory for structuring the processes evident in transcripts.

The inductive analysis provides an approach to the investigation of students' strategies and obstacles encountered in the reconstruction method, which are addressed in research questions (R1) and (R2) in Section 3.1. In analysing data for possible strategies, only the describer's utterances were considered, since the describer is the one who is dominantly shaping the actions and choosing how to solve the underlying task in the reconstruction method. However, both utterances of describers and builders are considered in order to establish possible obstacles arising in the reconstruction method, because the obstacles encountered arise within the communication processes between both students in the reconstruction method. Under consideration of the goals initiated by the research questions (R1) and (R2), the data coding was carried out based on a grounded theory analysis approach of open coding:

> During initial coding, incidents, events and items of interest are identified and labeled with code names that reflect a particular conceptual aspect of each of these phenomena. As analysis continues, codes having similar attributes are grouped together into categories representing a higher level of conceptual abstraction.
>
> Open coding starts as soon as the first set of data has been gathered. This coding consists of two analytic, meaning-making procedures, (1) asking questions of the data and (2) constantly comparing indicents. The goal of this process is to conceptualize the data into a collection of codified phenomena, or "substantive codes" that abstractly identify particular aspects of the empirical area study. (Teppo, 2015, p. 6)

A similar open coding was used in the pilot study of this present research, since strategies identified in previous studies about spatial abilities, such as analytic

118

and holistic strategies (cf. Barrat, 1953; see Section 2.1.5.2) were not considered suitable to code the conceptual aspects of the phenomena observed in the reconstruction method. One of the reason for the inadequacy in applying previously developed category systems for analysing strategies in spatial tasks is the lack of research in spatial-verbal tasks based on the reconstruction method in mathematics education research. As a starting point, the data coding in the pilot study served as basis for the futher development of codes and categories in the data analysis of the main study. However, the data coding process in the main study was more structured given the refinement of the research goals and the underlying intentions. The logic of abduction was used in the qualitative data analysis to construct new categories or sub-categories to describe and explain the empirical phenomena observed in the data analysis (cf. Kelle & Kluge, 2010).

Due to the two different foci on students' strategies and obstacles in the reconstruction method, two category systems needed to be developed: a category system for strategies and a category system for obstacles. The establishment of the categories to represent strategies or obstacles required refinement, optimisation, classification and restructuring of categories, which were mainly conducted based on two issues. The first issue was to establish which linguistic elements or phrases indicate a possible use of a particular strategy or obstacle, including considerations about the category it could be assigned to. The second issue was to identify text passages which show identical or similar characteristics in order to be assigned to the same category. If a text passage indicated the use of a previously non-categorised strategy or obstacle, then the researcher had to consider which categories or sub-categories had to be constructed in order to represent the "new" strategy or "new" obstacle. Therefore, the following questions were important for the ongoing process of identification and refinement of strategies or obstacle based on an interpretative-heurmeneutic approach (see Section 3.2):

1) *Which describer utterrances represent the use of a possible strategy?[33]*
2) *Which student utterrances represent failure or misinterpretation of communicated knowledge? When do describers reach their linguistic limit during the solving of the task in the reconstruction method?[34]*
3) *Which categories can be used to describe the patterns observed in text passages identified in 1) or 2)?*

---

[33]   Only applicable for sub-question (1) (see Section 3.1).
[34]   Only applicable for sub-question (2) (see Section 3.1).

4) *Which text passages show identical or similar characteristics and there-fore can be assigned to the same category or categories? If not, which categories or subcategories can be constructed during comparison and identification of patterns in the data analysis?*

Hence, identified categories have been refined and restructured after multiple phases of data coding, which lead to the establishment of sub-categories or neg-ligence of categories (in the case of duplication of categories or if a category did not really represent a strategy or obstacle) identified in previous phases of data coding in the main or in the pilot study. The established categories were labelled and described by their meaning, key characteristics, scopes and limitations (cf. Thomas, 2006). In the next step of the data coding process, theoretical considera-tions about conceptual ideas represented in the coded phenomena of the recon-struction method were considered for further development and discussion of codes and categories. According to Teppo (2015), this phase can be character-ised by the use of heuristic concepts or coding paradigm, which "provides a par-ticular theoretical perspective and set of heuristic concepts that structurally guide researchers as they begin to code for specific categories and identify relation-ships among categories" (Teppo, 2015, p. 10). This leads to the final phase of da-ta analysis in grounded theory research, which is the integration and further gen-eration of theory based on the coding of data collected in the reconstruction method (see Figure 21).

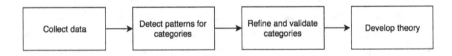

Figure 21 **Visualisation of the inductive data analysis process**

4.2.6.3 Deductive analysis

A deductive analysis is the analysis of data which tests "whether data are con-sistent with prior assumptions, theories, or hypotheses identified or constructed by an investigator" (Thomas, 2006, p. 238). Following a data-driven analysis in the inductive analysis, data collected in the reconstruction method was investi-gated quantitively using a pre-defined set of theoretical codes or hypotheses in order to reach the second level of interpretive understanding of how different students solve the tasks in the reconstruction method under consideration of research questions (R3) and (R4). In the deductive analysis, the analysed data consisted solely of the describers' utterances in the reconstruction method due to the consideration of describer's competencies and other factors for investigating hypotheses.

One of the pre-defined set of codes for the analysis of data is structured by the category system established for strategies in part of the the inductive analysis described in Section 4.2.6.2. As already described in Section 3.1, the frequency of occurrence of the identified strategies in the task in the reconstruction method was investigated for providing results for sub-question (3) using statistical devices (e.g., contingency tables and chi-squared tests). Due to the two-fold task embedded in the reconstruction method (see Section 4.2.2), the identical strategies[35] were only coded once, even if used on two different different spatial objects A and B or used more than once on the same object. Apart from investigating the frequency of use of identified strategies, generated hypotheses based on previous findings concerning the role of language proficiency, spatial abilities, and sex will be tested with the analysed data (see Figure 22).

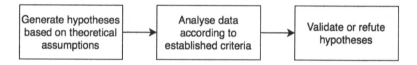

Figure 22 Visualisation of the deductive data analysis process

Similarly, the structural analysis of spatial language aimed at answering research question (R4) is based on pre-defined categories developed from theory and previous findings about spatial language. An analysis of the describing students' utterances in spatial discourse requires a quantification of aspects characterising spatial language and their analysis by using frequency methods. This approach of counting frequency of words in discourse was adopted from similar approaches in linguistics, whereby frequency of occurrence is considered as an important tool for analysing linguistic structure (cf. Diessel & Hilpert, 2016). Based on previous studies in cognitive psychology (e.g., Logan, 1988; Zacks & Hasher,

---

[35] Identical strategies will be considered as strategies which are reconstructed from identical or similar verbal utterances in spatial discourse, and used by the student for the same intended goal during the whole task solving process. In this case, the multiple use of the identical strategies should be considered as a repetition.

2002), Diessel & Hilpert (2016) state that frequency is an important factor for determining the level of knowledge acquisition and storage among children and adults and it "has a significant impact on sentence processing and utterance planning, and the development of linguistic structure in acquisition and change" (Diessel & Hilpert, 2016, p. 3). In the case of this study, it will be assumed that the frequency of use of particular elements or words which carry spatial meaning might be related to the student's language proficiency, spatial ability or sex (by formulating underlying hypothesis in Chapter 6), which are important factors for solving the spatial-verbal tasks in this study. Due to the different lengths of describers' spatial discourses in the reconstruction method, the relative frequency of linguistic components of spatial language, which is a percentage made up of the ratio of actual occurrences of these components to the total number of phrases in the student's spatial discourse, was considered. In this approach, a phrase was considered as a set of words which form a conceptual unit and a clause. Therefore, the student's utterances in the transcriptions needed to be additionally broken down in phrases for the deductive analysis. Statistical hypothesis tests, such as the nonparametric test Mann-Whitney U Test were used to test any significant difference concerning the hypotheses generated from previous research findings concerning the use of spatial language and its dependency from other factors, such as students' language proficiency, spatial abilities, and sex (see Figure 22).

Thus, the following questions play an important role in the deductive analysis:

1) *Which strategies have been used and how often by which students based on the dichotomies of language proficiency, spatial abilities, and sex?*[36]

2) *Which components of spatial language differ among their use by different students based on the three dichotomies?*[37]

3) *Which hypotheses can be generated about the use of identified strategies under consideration of students' language proficiency, spatial abilities, and sex?*

4) *How can data be represented in order to support or reject the hypotheses generated in 3)?*

---

[36]   Applicable only for research question (R3) (see Section 3.1)
[37]   Applicable only for research question (R4) (see Section 3.1)

4.2.6.4 Presentation of results

As it has already been pointed out in the previous sections, the data analysis consisted of four cycles, two of inductive, and two of deductive analyses. Both approaches of data analysis should contribute to a better understanding of how students solve spatial tasks using spatial language. Whilst the categories identified and described in the inductive analysis reveal more about the nature of the strategies or obstacles encountered in the solving process, the development and adaptation of coding systems to test hypotheses about the use of the previously identified strategies and the structure of spatial language concerning students with different background provides a more explanatory approach to the research goals of this present study. Moreoever, the second focus in the deductive analysis part presumeably presents a first approach to analysing spatial language structurally and differentiates among its use by different students in mathematics education.

The results in the inductive data analysis part are described and illustrated by referring to excerpts from the transcripts of the reconstruction method. Transcripts are selected according to their adequacy and their comprehensiveness concerning the underlying identified strategy or obstacle, if applicable, supported by visualisations of intended object parts (by the describer) or the reconstructed object (by the builder). For a better understanding of the quantitative results in the deductive analysis part, the results are represented using bar charts or numeric values in statistical hypothesis tests and then used to test the pre-described hypotheses. The results of the whole data analysis process of the main study are reported in the next chapters: the results from the inductive data analysis in Chapter 5 and the results from the deductive data analysis in Chapter 6.

5. Results and discussion from the inductive data analyses

In this chapter, the main results from the data analysis in the inductive data analyses of the present study are presented. The presentation of the results corresponds to the data analysis process described in Section 4.2.6. In Section 5.1, the identified strategies are described, followed by a description of obstacles encountered by the students in solving the spatial tasks in the reconstruction method in Section 5.2.

5.1 Description of the identified strategies

To answer the first research question (R1), I first present the types of strategies students used when describing spatial configurations in the spatial tasks. In this study, strategies in spatial tasks refer to actions chosen by a subject to reach the goal in the spatial task, independent of whether the use of the strategy is successful or not (see Section 2.1.5.1). Moreover, the identified strategies account for the students' individuality represented in his or her cognitive processes in the solving of the spatial task in the reconstruction method. The findings from this present study reveal six main strategies used by students when solving spatial tasks: *spatial metaphors, object break-down, assembling, rotation, cubes controlling*, and *structure controlling strategy*. In the following sections, the identified strategies are described and explained with reference to theoretical foundations.

5.1.1 Spatial metaphors

The first identified strategy, which is referred to as *spatial metaphors* (M), was one of the most common strategies during the inductive data analysis of the students' discourse during the solving of the spatial tasks in the reconstruction method. *Spatial metaphors* are characterised by the use of metaphors to communicate and describe spatial knowledge activated when handling and describing spatial objects. Following Lakoff and Núñez (2000), metaphors are projections from source domains to target domains, in which the source's properties and characteristics are assigned to the target (see Section 2.2.5). In the case of spatial metaphors, properties from different sources (e.g., everyday life phenomena) are transferred to the whole or part of a spatial object.

Spatial metaphors are considered as strategies due to their manifoldness and possibility of choice in use among students solving the task in the reconstruction method. A deeper investigation of the diverse use of spatial metaphors highlighted the need of categorisation of different spatial metaphors used by students during the reconstruction method according to different aspects. This analysis of spatial metaphors led to the establishment of three different dimensions of spatial metaphors – the linguistic, the spatial and the conceptual dimension of spatial metaphors – which are explained in the following sub-sections.

5.1.1.1 The linguistic dimension

From a linguistic point of view, spatial metaphors can be expressed using linguistic elements from different levels of language acquisition (see Section 2.2.4). In order to represent the different facets of the language used in different spatial metaphors, the following distinction wasestablished: everyday, letter-based, and mathematical spatial metaphors (see Figure 23).

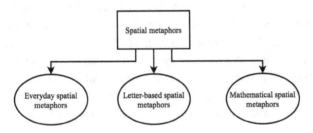

Figure 23 Categorisation of spatial metaphors along the linguistic dimension

From a theoretical perspective (for example, traditional definitions for classification or assignment of terms and concepts to everyday or mathematics language) and for feasibility in data analysis processes, these categories were considered as distinct. The categorisation of spatial metaphors in the three categories is justified by the following steps in spatial knowledge acquisition: students learn everyday language about space primarily outside school, during school time they develop their everyday language and increasingly use symbols and characters for representing mathematical knowledge, in particular, spatial concepts, and further develop their mathematical spatial language in geometry lessons. Another important issue regarding the distinction between student everyday and mathematical language in the data analysis process was the consideration of the students' prior knowledge or learning experience for establishing an appropriate concept of mathematical language. One of the underlying assumptions about mathematical language in this present research is that since mathematical language develops through learning processes and increasing exposure time, a flexible[38] student-driven notion of mathematical language should be considered.[39] Establishing what might be assigned to mathematical language for secondary school students requires not only theoretical definitions, but also linking the students' use of mathematical concepts to the expectations in real implementation, in this case to the learning experience of fifth grade students.[40] In the following sub-sections, everyday, letter-based and mathematics spatial metaphors are described in detail.

5.1.1.1.1 Everyday spatial metaphors

An *everyday language* or *everyday spatial metaphor* is a metaphor used by the students to describe spatial characteristics by drawing on linguistic means from everyday language in the spatial task solving process. When using *everyday spatial metaphors*, students do not use technical words from the mathematical language, but instead they use illustrative ideas and images from the real world to describe a spatial configuration in space using everyday language. Therefore, *everyday spatial metaphors* are projections, in which the source domain originates from phenomeona in everyday life or surrounding environment and the target domain is a particular spatial configuration in the spatial task in the reconstruction method. When using *everyday spatial metaphors*, students transfer the properties of an everyday object or situation to the spatial object or its internal part (see Table 16). Since everyday spatial metaphors relate a target domain within mathematics (geometry and spatial abilities) to a source domain outside of mathematics (everyday-life), these metaphors are considered as grounding metaphors according to Lakoff and Núñez (2000) (see Section 2.2.5).

---

[38] The flexibility of mathematical language refers to the issue that what belongs to mathematics language is not necessarily static or pre-defined, but it can change depending upon the student and his previous experience in mathematics learning or teaching.

[39] In other words, the mathematics language of students at early secondary school is different than mathematics language of students at the end of secondary school, as the mathematics language of students at university level.

[40] The emphasis is on the importance of implementation in mathematics classroom guided by the question: *Which words or terms do we expect students to use in mathematics classroom in order to recognise a development in their language?* The teacher's expectations concerning this issue should be linked to the classes and learning experience of students.

Source domain	→	Target domain
Properties of an everyday phenomenon	→	Spatial properties of object

Table 16   Mapping in everyday spatial metaphors

The use of this strategy requires students to activate their previous everyday experience and generate corresponding mental images when acting on the spatial-geometical object, and use this knowledge to describe a particular spatial configuration and communicate the underlying spatial knowledge. The following are two examples of students' everyday spatial metaphors which are used by the students to describe the underlying spatial knowledge of an object or its internal part. The first transcript excerpt, Transcript 1, shows the use of everyday spatial metaphor *staircase*, which is used to describe spatial object A in the reconstruction method.

Transcript 1:

Describer 2: *Okay, at the bottom it has... firstly it has only one... here... simply only like a balk with four stones and to it a staircase downwards with three, two and one.*

Okay, es hat unten... erstmal hat es nur eine... hier... einfach wie nur ein Balken mit vier Steinen und da dran eine Treppe nach unten mit drei, zwei und eins.

Figure 24   Intended internal part of spatial object A in Transcript 1

Describer 2 associates spatial object A with the notion *staircase* from everyday context. The structural property of a prototypical staircase, the bridging of a vertical distance divided into smaller equidistant vertical distances, is transferred from the source domain to the intended internal part of the spatial object visualised in Figure 24. Another example of an everyday spatial metaphor is illustrated in the following transcript of another describer's utterances during spatial task A in the reconstruction method:

128

Transcript 2:

(Describer 4 gets a glimpse of the following builder's object)

(Describer 4 continues with his description of spatial object A)

	(...)	(...)
Describer 4:	*And... uhm... they are two foursome towers, right?*	*Und... ähm... die sind auch zwei Vierertürme ne?*
Builder 4:	*Yes.*	*Ja.*
Describer 4:	*And uhm... there at the three-some next to it.*	*Und ähm... da bei den Dreier daneben.*
Builder 4:	*Yes.*	*Ja.*
Describer 4:	*You can take these off please and then at the foursome tower there... uhm...do it like a curve then just to it.*	*Die kannst du bitte abmachen und dann bei den Viererturm da... ähm... so wie so ne Kurve dann halt dranmachen.*
	(...)	*(...)*

Describe 4 uses two everyday spatial metaphors in her speech: *tower(s)* and *curve*. Both spatial metaphors *tower* and *curve* orginate from everyday contexts, and assign the property of being high in proportion to its lateral dimensions and the property of a non-zero curvature to the spatial object A in the reconstruction method respectively.

5.1.1.1.2 Letter-based spatial metaphors

The inductive analysis of spatial language used for spatial metaphors in the re-construction method led to the identification of another class of spatial meta-phors: *symbol-based* or *letter-based spatial metaphors. Letter-based spatial met-aphors* are spatial metaphors which are realised by the use of symbols or charac-ters from the written form of language, for example, Latin alphabet letters[41], in

---

[41] Due to the use of different symbols and characters representing written languages world-wide, this category is highly culturally (and educationally) shaped. For instance, Greek or Arabic alphabet can activate different spatial knowledge, which cannot be realised using Latin alphabet characters or vice versa. For example, the Greek letter *delta* Δ in uppercase can be used to represent the spatial-geometrical characteristics of forms of a triangle. But a high number of characters in some Arabic scripts, which tend to use lot of curvatures in

spatial discourse. The use of such symbols or characters activate more specific spatial knowledge reconstructed from the conventionally used symbols or characters which are part of the alphabet. Although the symbols and characters in the alphabet are also present in everyday life, in contrast to everyday metaphors, this category makes use of a limited pool of speech elements based on written form of language (i.e. symbols and characters of alphabet) which is used for communicating more specific spatial knowledge. Therefore, letter-based spatial metaphors appear to be characterised by a higher abstraction in terms of spatial configurations, whereby the spatial relations between lines constructing a symbol are transferred to the spatial object (see Table 17).

Source domain	➜	Target domain
Properties of a symbol or character from written language	➜	Properties of spatial object

Table 17   Mapping in letter-based spatial metaphors

In particular, letter-based spatial metaphors were frequently used by the students describing spatial object B, as the following transcript excerpt shows:

Transcript 3:

Describer 15: *Well it looks like an H, well it is so to say an H, only that the H... well it is built the same as an H.*

Also es sieht aus wie ein H, also es ist sozusagen ein H, nur dass das H... also es wird genauso gebaut wie in H.

Builder 15: *How high is the H?*

Wie hoch ist das H?

Describer 15: *It has five cubes at the right and at the left. And it has two times cube, well two cubes are next to each other. (...)*

Es hat an rechten und an der linken fünf Würfel. Und es hat halt zweimal Würfel, also zwei Würfel sind nebeneinander. (...)

Describer 15: *It is quite a fat H. Well there are two Hs. It has one, two, three, four, five cubes at the right and at the left, and one cube in the middle. (...)*

Das ist ein ganz dickes H. Also zwei Hs. Eins, zwei, drei, vier, fünf Würfel hat es an der rechten und an der linken, und in der Mitte ist ein Würfel. (...)

---

writing, are less ideal for representing, for instance, certain pre-defined spatial relations, for example, orthogonal constellations (the closest being the symbols *laam* ل or *daal* د). This is another reason for differentiating between this category and the category of everyday spatial metaphors.

In Transcript 3, the dominant letter-based spatial metaphor H[42] is the describer's first association with the symmetrical spatial object, which was one of the underlying considerations for the object design (see Section 3.5.2.4). The use of the letter-based spatial metaphor activates a particular pre-defined spatial-geometrical constellation of three lines (two vertical lines and a connecting orthogonal line inbetween, which are referred to as *right*, *left* and *middle* in Transcript 3) and the underlying symmetrical characteristics, which are transferred on the spatial object itself. At the end of Transcript 3, the student's description of the spatial object as two Hs reveals the way how multiple use of letter-based metaphors can be used to describe layers of spatial object. Whilst being predominantly used in spatial task B, letter-based metaphors were also used by describers in spatial task A, as it is apparent in the following transcript excerpt:

Transcript 4:
(The researcher shows Describer 6 the reconstructed object and instructs the describer to optimise his description)

Describer 6:	Yes, she has done it correct like that. But she should do like this and this together.	Ja, so hat sie schon richtig gemacht. Aber die sollte dann so und so zusammen machen.
Researcher:	Exactly, you must describe it to her somehow.	Genau, das musst du ihr jetzt beschreiben irgendwie.
Describer 6:	Yes, then... yes the... I have the whole time, like a T at the end. Almost like a T. (the describer uses gestures and actions on spatial object to show how the parts shoud be linked together)	Ja dann... ja das... hab ich schon die ganze Zeit, wie ein T am Ende. So fast wie ein T. (der Beschreibende benutzt Gesten und Handlungen um zu zeigen, wie die Teile zusammengesetzt werden sollen)
Builder 6:	(puzzled) Like a T? Like a T at the end. (...)	(verwirrt) Wie ein T? Wie ein T am Ende. (...)

---

[42] During the reconstruction method, the intended use of letter-based metaphors in object B was of capital letters, whereby mostly the capital was not explicitly mentioned by students in their spatial discourse.

Transcript 4 illustrates the use of the letter-based spatial metaphor T, which was introduced by Describer 6 to explain how to connect the two internal parts of the objects correctly, after taking a glimpse at the reconstructed object. The use of this spatial metaphor was intended to transfer the spatial-geometrical constellation between the two lines (one vertical and another one horizontal on top of the first one) represented by the underlying symbol to the orthogonal spatial relation between the two internal parts of the spatial object.

5.1.1.1.3 Mathematical spatial metaphors

Mathematical spatial metaphors are metaphors whose linguistic elements originate from mathematics language[43], and which is used by students to solve the spatial task in the reconstruction method to describe spatial characteristics. When using *mathematical spatial metaphors*, students use terms from the mathematical language to describe a spatial configuration. Thus, *mathematics spatial metaphors* can be described as projections from other domains of mathematics (non-spatial) in the source domain to a spatial configuration in the target domain. Hereby, properties from mathematical concepts developed in a presumably non-spatial context (e.g., two-dimensional geometrical concepts) are transferred to spatial objects (see Table 18).

Source domain	→	Target domain
Properties of a mathematical concept which is not necessarily of spatial nature	→	Spatial properties of object

Table 18 Mapping in mathematics spatial metaphors

In *mathematics spatial metaphors*, both source and target domain originate within mathematics, the latter in the domain of spatial abilities in mathematics educa-

---

43 As already described earlier in this section, the assignment in mathematics and everyday language is dependent on the mathematics learning experience of the subject. In the case of this present study, the students were attending the fifth grade and still developing their mathematical language. Therefore, one important issue for considering which metaphors belong to mathematics spatial metaphors was the potential use of the corresponding terms in mathematics lessons till the end of the fifth grade. Moreover, student mathematics language can be described as showing linguistic patterns similar to the teacher's mathematics language. Therefore, the notion of mathematical language is rather student-oriented in this present study.

tion (in terms of Pinkernell, 2003; see Section 2.1.3.4). Therefore, *mathematical spatial metaphors* can be considered as linking metaphors according to Lakoff and Núñez (2000) (see Section 2.2.5).

The use of mathematics spatial metaphors requires the students to activate knowledge presumeably learnt in previous mathematical learning processes. Under consideration of the research goals of this present research, the focus in analysing mathematics spatial metaphors should not be whether the students' conceptual development is flawless or not, but the emphasis is on the analysis of metaphors used to understand students thinking during the solving of the spatial task. Another issue to address in this category is the extent to which mathematical concepts or geometrical concepts are considered as non-spatial. In this case, the students' intended meaning should be considered in order to perceive the intended spatialness of the concept, since students are likely to use mathematical concepts with other meanings rather than the conventional meanings as defined in traditional mathematics. If such a difference in meanings is established, then its use is categorised as a mathematical spatial metaphor.

The following transcript excerpt, Transcript 5, illustrates an example of a mathematical spatial metaphor by a student describing spatial object A in the reconstruction method:

Transcript 5:
(Describer 15 gets a glimpse of the builder's object)

(Afterwards the Describer 15 continues with his description of object A)

Describer 15:	*Well, it was pretty good. Well do not change anything, only... uhm... you have done both of them straight together, right?*	*Also das war schon ganz richtig. Also nichts mehr umändern, nur... ähm... du hast die beiden so gerade zusammengemacht, oder?*
Builder 15:	*Yes.*	*Ja.*
Describer 15:	*And make them so together, that it looks like a correct triangle.*	*Und mach die mal so zusammen, sodass wie so ein richtiges Dreieck aussieht.*
	(...)	*(...)*

In Transcript 5, the describer uses the mathematical spatial metaphor *triangle* for describing how to connect the two internal parts of the object together. The student's notion of *triangle* is used based on a specific type of triangle, which is emphasised by the word *richtiges* [*correct*] in the last utterance of the transcript excerpt. Due to the intended spatial constellation between the two parts of the object in the description, the student presumably refers to a right-angled triangle. Therefore, *triangle* should be considered as a mathematical spatial metaphor, whereby the source domain is geometry, since geometrical characteristics of the specific triangle are transferred to the spatial object(s) in the task solving process.

The importance of the student's intended meaning in spatial discourse, which is reconstructed by the researcher in the analysis of spatial language, shows how even mathematical concepts of spatial nature can be used metaphorically. Consider, for instance, the following transcript excerpt of a student describing spatial object A in the reconstruction method:

Transcript 6:

(Describer 9 attempts to describe to the builder how to build spatial object A again, after having taken a look at the reconstructed object)

	(...)	*(...)*
Describer 9:	*Do two double and two single.*	*Mach zwei Zweier und zwei Einer.*
Builder 9:	*Ok. Next to each other*	*Okay. Die auch nebeneinander?*
Describer 9:	*No...Yes well like a pyramid, now you do one high (points his finger to left most cube and moves it to the highest point of the spatial object) only at the top... the two foursome*	*Nein... doch so wie eine Pyramide machst du jetzt eins hoch (zeigt mit seinem Finger auf den untersten linksten Steckwürfel und bewegt ihn zur höchsten Stelle des Objektes) nur oben... an der Spitze sind die zwei Vierertürme.*
Builder 9:	*I have the foursome towers now and then still the threesome.*	*Ich habe jetzt die Vierertürme und dann noch die Dreier.*
Describer 9:	*And then after the threesome, the double.*	*Und dann nach den Dreiern die Zweier.*
Builder 9:	*Yes, I have done it.*	*Ja, habe ich gemacht.*
Describer 9:	*And then still the double, a single.*	*Und dann noch den Zweier, ein Einer.*

Builder 9:   *Well build the double left*   *Also links und rechts das Zweier*
               and right at it and then the   *dran bauen und dann die Einser,*
               singles, right?   *ja?*
Describer 9:   *Wait. Do you have some-*   *Warte. Hast du was Ähnliches wie*
               thing similar like a pyra-   *eine Pyramide?*
               (...)   *(...)*

In the above utterances, Describer 9 uses *pyramid* to describe the overall structure of the spatial object, once the internal parts were connected to each other as
instructed. In his description, Describer 9 seems to reduce the underlying spatial
object to some characteristics, for example, having a tip, being symmetrical and
a diagonal connection in space between the left-most cube at the bottom (as indicated by the describer's gestures in Transcript 6). It can be inferred that Describer 9 chose to describe the given object using the term *pyramid* because the object's highest point was in the middle, and its left-most and right-most cubes
were at the lowest position. Adopting an interpretative approach, it appears that
Describer 9 invokes a mental image of a two-dimensional front view of a pyramid, which is equivalent to a prototypical isosceles triangle, although the term
pyramid refers to a three-dimensional object. Therefore, the use of *pyramid* denotes a mathematical spatial metaphor, which induces the transfer of the property
of convergence to single point at the top on the underlying structure of the spatial
object.

Whilst analysing the different metaphors from a linguistic perspective, it was noticed that metaphors are not only expressed using different linguistic means, but
are also intended for different purposes. The different purposes for which spatial
metaphors were used in the task of the reconstruction method are discussed in
the next section.

5.1.1.2 The spatial dimension

Students solving spatial tasks in the reconstruction method used spatial metaphors for different aims or purposes, which are also referred to as functions of
spatial metaphors. The functions of spatial metaphors point out the different spa-
tial knowledge which is communicated by using spatial metaphors, hence consti-
tuting the spatial dimension of spatial metaphors. Three functions of spatial metaphors were identified: *Structure (S), Spatial Position (SP)*, and *Spatial Relation
(SR)*. The function *Structure* denotes the use of spatial metaphors for describing
the structure of the object or its internal part. *Spatial Position* refers to spatial

metaphors which are used to describe the position or spatial orientation of a spatial object in space. The spatial position of an object encompasses spatial information of where and how the spatial object is standing in space (e.g., vertical, horizontal and oblique perception) under consideration of the subject's own body, relatively independent from the position of other spatial objects in space[44]. The function *Spatial Relation* refers to the use of spatial metaphors to describe the spatial relation between two or more objects in space (see Section 2.1.2.1). In the case of spatial metaphors, the spatial relation function should refer to the spatial relation between at least two objects excluding the individual's body. Consider, for instance, the following transcript excerpt, in which everyday spatial metaphors are used to describe the structure and spatial position of objects:

Transcript 7:

(Describer 8 tries to describe how the object should be positioned in space)

Describer 8:	*No, I mean it should... the foursome staircase should uhm... well the tip should [come] to you... look in your direction, like this... therefore straight... not to the left*	*Nein, ich meine die soll... die Vierertreppe soll ähm... so die Spitze soll zu dir... so zu dir gucken... also gerade... nicht nach links oder nach rechts.*
Builder 8:	*Yes, I haven't done that, it is straight upwards.*	*Ja, habe ich ja nicht, die ist grade nach oben.*
Describer 8:	*No not upwards... like this downwards, so... so that it looks like a staircase which one can go up.*	*Nicht nach oben... so nach unten, dass... dass das so aussieht wie ne Treppe ist wie man da hochgeht.*
Builder 8:	*Yes, now it is like that.*	*Ja, ist ja jetzt so.*
Describer 8:	*Yes, and this should then come straight to you.*	*Ja, und das soll dann so gerade zu dir kommen.*
Builder 8:	*This means straight to me... this means...*	*Das heißt gerade zu mir... das heißt...*
Describer 8:	*It should look at you!*	*Das soll zu dir gucken!*

---

[44]   In contrast to the definition of Frostig et al. (1977), spatial position should not be understood merely as the subject's ability to identify and perceive the position of an object in space under consideration of his own body. In the inductive analysis of the present study, spatial positions of the object should be additionally perceived as the ability to describe how a spatial object is standing in space possibly by itself, for example, whether it is vertical or horizontal, or explicitly in relation to the subject's body.

| Builder 8: | This means that I can practically see through it there... so to say. | Das heißt, dass ich da so praktisch durchgucken kann... sozusagen. |
| Describer 8: | Well, that the staircase uhm... looks at your sweater like in... | Na ja, dass das halt die Treppe ähm... zu deinem Pullover guckt so wies... |

In Transcript 7, Describer 8 uses the metaphor *looking* for describing the position of the spatial object referred to as *staircase* in the reconstruction method. In her description, Describer 8 assigns properties of human beings to the spatial objects, such as the ability of looking or seeing, in order to place the object in a particular direction in space. In addition, Describer 8 uses the idea of 'going up the spatial object' as a spatial metaphor to position an object in space. Whereas the *staircase* metaphor is a spatial metaphor for describing the structure of the internal part of the spatial object, the *going up* spatial metaphor denotes an action which is metaphorically performed by the subject on the spatial object in order to describe its position in space under the consideration of his or her own body position.[45] An example of spatial metaphor with spatial relation function can be found in the following transcript (which is a continuation of Transcript 1):

Transcript 8:

| Describer 2: | The balk with four and then there is the staircase three, two, one. And then you must... you have a wall and then you must do the four, three, two staircase to the right. | Der Balken mit vier und dann kommt die Treppe drei, zwei, eins. Und dann musst du... hast du ja die Mauer und dann musst du nach rechts die vier, drei, zwei Treppe machen. |

In Transcript 8, Describer 2 uses the everyday spatial metaphor *wall* to describe the spatial relation between the two internal parts of the spatial object (see Figure 25). The spatial metaphor *wall* does not describe the structure of the second internal part of the object, since this has a similar structure as the first one (i.e. *a balk with stairs*). In his description, the student uses the spatial metaphor *wall*

---

[45] In this case, the act of walking on the spatial object is considered as a metaphor with SP-function, since the subject's body and his imaginary actions are used to communicate knowledge about the spatial object in space, hence that the spatial object should be constructed upwards. Although argueably an extent of spatial knowledge about the structure is also communicated by the *going up* spatial metaphor, the main metaphor for structuring the spatial object is the spatial metaphor *staircase*, which is used in the same utterance by the describer.

when moving from the description of the first internal part to the second one. The spatial metaphor *wall* is used to denote the spatial constellation of the two parts in relation to each other, which resembles a situation in which two walls meet each other to create a corner. Hence one of the internal objects is acting as a *wall* in relation to the other one, as it can noticed in the following utterance, *"And then (...) you have a wall"*. In this case, the property of two adjacent walls set at a 90 Degrees angle is transferred from a real-life context of a wall in a room to describe the spatial relation between the two internal parts of the spatial object A. Hence, *wall* is an example of an everday spatial metaphor with SR-function. An example of a mathematical spatial metaphor with SR-function is used by the same describer, Describer 2, in previous utterrances during the reconstruction method (see Transcript 1 for spatial metaphor *triangle*).

The mapping of properties of the exemplary spatial metaphors mentioned above under consideration of their underlying function in spatial discourse are presented in Table 19.

Source domain	→	Target domain
Living organism	→	Spatial object
Ability of looking	→	Spatial position / orientation of an internal part of object
Staircase	→	Spatial object
Property between a number of stairs	→	Structure of an internal part of object
Wall	→	Spatial object
Property of two adjacent walls	→	Spatial relation between two internal parts of the object

Table 19   Mappings in exemplary spatial metaphors with SP-function, S-function and SR-function respectively

5.1.1.3 The conceptual dimension

Apart from the linguistic repertoire and the intended functions in spatial context, the third categorisation of spatial metaphors is according to the nature of the underlying conceptions, which is referred to as conceptual dimension of spatial metaphors. In the theoretical background about representation of mathematical concepts (see Section 2.4.1), Sfard's (1999) theorical framework about the dual nature of mathematical concepts was introduced. Mathematical concepts can be developed in two conceptions, in structural and operational, which have been described respectively as following:

Seeing a mathematical entity as an object means being capable of referring to it as if it was a real thing – a static structure, existing somewhere in space and time. It also means being able to recognize the idea 'at a glance' and to manipulate it as a whole, without going into details. (Sfard, 1991, p. 4). (in Section 2.4.1, p. 63)

In contrast, interpreting a notion as a process implies regarding it as a potential rather than actual entity, which comes into existence upon request in a sequence of actions. Thus, (…) the operational [conception] is dynamic, sequential, and detailed" (Sfard, 1991, p. 4). (in Section 2.4.1, p. 63)

A similar categorisation was noticed among the spatial metaphors used among the describing students for activating particular spatial concepts in the discourse. Whereas some spatial metaphors referred to spatial concepts of a static nature, which do not necessarily require any actions or movement in space, other spatial metaphors activated a sequence of actions on the spatial object in space, hence being of dynamic nature. Therefore, the two conceptions evoked by the different spatial metaphors are referred to as static and dynamic conceptions accordingly. Consider, for instance, the following transcript, in which the student uses the spatial metaphors *staircase* and *walking up* or *down* in the description of spatial object A:

Transcript 9:

(Describer 12 explained how two parts of the objects should be constructed, and attempts to describe how the overall structure of the object is)

| Describer 12: | *And now do these three steps to the other staircase, which you have made. Uhm… place it infront of you in such a way as if you would… must walk up… and then… and now take the other staircase which you have just made and do it uhm… at the most front… as if one at the front uhm… walks up at the front and then uhm…* | *Und jetzt mach diese drei Stufen an die andere Treppe, die du gemacht hast. Ähm… stell die mal vor dich so als würdest… müsstests du hochlaufen… und dann… und jetzt nimm die andere Treppe die du jetzt grad gemacht hast und dann pack die ähm… ganz vorne ran… so als wenn man vorn äh… von vorne hochläuft und dann äh… die andere Treppe wieder runtergeht.* |
| | *(…)* | *(…)* |

In Transcript 9, the spatial metaphor *staircase* represents a spatial metaphor of static nature, whereby the metaphor represents the underlying spatial concept of the internal part of the object wholly. In contrast, the spatial metaphor of walking up or down on the spatial parts of the object is a dynamic spatial metaphor, because it induces actions performed on the spatial object which reveal more about the spatial details which are not necessarily addressed by the static spatial metaphor *staircase*. Further examples of static spatial metaphors are *curve* and *pyramid* in Transcript 2 and 6 respectively, and an example of dynamic spatial metaphor is *looking* in Transcript 7.

5.1.1.4 Overview of spatial metaphors

As described in the preceding sections, there are three categories of spatial metaphors used as strategies by students to solve spatial tasks in the reconstruction method. The first categorisation, the linguistic dimension, is based on the linguistic elements used for spatial metaphors and includes three strands: *everyday* (E), *letter-based* (L) and *mathematical* (M) spatial metaphors. The second content-based categorisation, the spatial dimension, was based on the spatial meaning of function of the used spatial metaphors. There are three different spatial aspects for which spatial metaphors are used: *structure* (ST), *spatial position* (SP) and *spatial relations* (SR). The third categorisation of spatial metaphors is in *static* (S) and *dynamic* (D) conceptions. The various spatial metaphors can be categorised using the three-dimensional model illustrated in Figure 25.

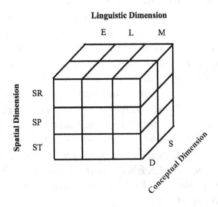

Figure 25 A model for the classification of spatial metaphors according to the three dimensions (linguistic, spatial and conceptual)

The model for spatial metaphors in Figure 25 can be used to localise and categorise spatial metaphors regarding their language use, spatial meaning and their conceptions. Spatial metaphors can be assigned to the underlying cubes in Figure 25, which visualise the intersections between the dimensions in the model. For instance, an example of spatial metaphor for the cube (M, ST, S) would be the static everyday spatial metaphor with ST-function *pyramid* in Transcript 6 and an example for spatial metaphors in cube (E, SP, D) is *walking up* and *down* in Transcript 9. However, the assignment of spatial metaphor to cubes should be regarded as an orientation, and depending on the strictness of definition of the underlying categories, an assignment to multiple cubes could be possible[46]. Consider, for instance, the spatial metaphor of *walking up* and *down* in Transcript 9. Whereas this metaphor is undeniably generated from everyday language and has a dynamic nature, it's spatial meaning can be interpreted as spatial position (since the student's body plays an important role for reconstructing the underlying spatial knowledge) or for emphasising the structure of the *staircase*, independent from the orientation (whether from down to top or from top to bottom). Hence, depending on the underlying emphasised spatial meaning, the spatial metaphor can be assigned to (E, SP, D) or (E, ST, D) respectively. Nevertheless, the model for spatial metaphors provides a basis for representing the different linguistic, cognitive and conceptual aspects in spatial metaphors. As described earlier in Section 5.1.1.1, the different categories of dimensions should not be regarded as entirely separate from each other, but rather as describing different emphases of context-embedded phenomena which can be reconstructed from analysis of spatial metaphors. With regard to its generalisation, this model offers only a representation of spatial metaphors among students at lower secondary level. The student perspective of spatial metaphors is especially visible on the linguistic dimension, hence, an application of this model to spatial metaphor use among other groups (e.g. mathematics teachers or mathematicians) primarily requires changes to the linguistic dimension, but not necessarily to the conceptual or spatial dimensions in the model.

---

[46] The classification of spatial metaphors in sub-categories illustrated in Figure 25 may possibly vary according to context, the subjects' intended goal and the researcher's subjective interpretations of knowledge. Therefore, the assignment of spatial metaphors to a particular sub-category should represent an orientation or tendency based on subject interpretation rather than just the unequivocal objectivity.

5.1.2 Object break-down strategy

In the strategy of object break-down (BD), describers mentally break down the object into several internal parts to reduce the complexity of the object's structure and facilitate the description of the intended spatial object in the spatial task of the reconstruction method. The breakdown of the spatial object is described explicitly by the describer in his or her spatial discourse and may be additionally supported by the use of gestures during the description. Whilst there are several ways of breaking down an object, in general, one has to consider that not all of these possibilities are helpful to facilitate the describing process. In this present study, the use of strategy object break-down is denoted as the breaking down of the object into different internal parts (based on objects A and B), each consisting of at least two cubes. For instance, if spatial object A is broken down into twenty cubes (which is the maximum number of cubes for spatial object A) in the spatial discourse, this was not coded as a breakdown strategy, since the construction of spatial object A using twenty cubes is more of a task demand (and less of a choice, which is characteristic for strategies).

On a spatial-linguistic level, the describers' breakdown of spatial objects can be done in several ways, for instance, by referring to the internal parts using different names, which is a characteristic for the object description in students' spatial language (see Section 2.3.1.1). The invention or assignment of names or nominal phrases for describing the spatial objects is also a typical characteristic of spatial language in the reconstruction method, which serves not only to describe the spatial object, but also to identify a particular part in the underlying spatial context. Examples of such names or phrases for spatial object parts are *part one* and *part two* or (situational invention of names or phrases) or *the bigger part* and *the smaller part* etc., or by using spatial metaphors with ST-function in phrases, for example, *the smaller staircase* and *the bigger staircase*. The giving of names for internal parts of the object may enable students to reduce ambiguity when referring to different parts in which the object has been broken down into for structuring space in their discourse.

There are different ways by which students can break down the spatial objects in the spatial tasks of the reconstruction method. A possible approach is by searching for congruent and similar[47] structures in the object (see Figure 26, Figure 28, Figure 31 and Figure 33). Another way to describe the contruction of a spatial object or its part is by perceiving the object or its part in terms of columns or rows (see Figures 29, 30 and 32). With the aim of facilitating the description and reconstruction of component parts of a given spatial object, students using these approaches often refer to similarities, regularities, and patterns in the structure of the object or its parts. Instead of describing the construction of the object as a whole, students describe how the internal parts of the object can be constructed separately. A characteristic of object breakdown is a statement about the number of parts in which the object is broken into, which should structure the problem solving process, as indicated in Transcript 10.

Transcript 10:

	(...)	*(...)*
Describer 6:	*Simply do the first staircase, then you must still do another staircase.*	*Mach einfach die erste Treppe, dann must du noch 'ne andere Treppe machen.*
Builder 6:	*Does it have to be two times?*	*Müssen die zweimal ankom-*
Describer 6:	*Yes, two staircases.*	*Ja, zwei Treppen.*
Builder 6:	*I cannot understand you... I already have one staircase.*	*Ich verstehe dich nicht... eine Treppe hab ich schon mal.*
Describer 6:	*Yes, you already have one staircase, then you must do another one.*	*Ja, eine Treppe hast du schon, dann musst du noch eine machen.*
Builder 6:	*Next to it?*	*Daneben?*
Describer 6:	*Four, three, two, one.*	*Vier, drei, zwei, eins.*
Builder 6:	*Next to it?*	*Daneben?*
Describer 6:	*Not next to it, just do the staircase first.*	*Nicht daneben, mach einfach die Treppe als Allererstes.*
	(...)	*(...)*

In Transcript 10, the describer breaks down spatial object A into two parts, each of which is referred to as *staircase*. Describer 6's break-down was emphasised by the sequential discourse structuring phrases indicating the actions of buildings two staircases, which were referred to as *first staircase* and *another staircase*.

---

[47]   The meaning of notion of similar in this context should not necessarily be equivalent to its meaning in mathematical language. Similar spatial configurations are configurations which are similar concerning to the way they are constructed using manipulatives.

Based on an analysis of spatial discourse, the researcher can reconstruct which break-down (see Figure 28) is intended by Describer 6, who describes the construction of the second part merely as *four, three, two, one* (see Figure 29 or 30). This shows the breakdown of the spatial object in congruent or similar parts, which facilitates the describer's description by activating previous construction patterns. Another example of breakdown, in this case of spatial object B, is illustrated in Transcript 3. In Transcript 3, Describer 15 breaks down the spatial object in two Hs, i.e. in two congruent parts, each of which is described using the letter-based spatial metaphors H, as the break-down visualised in Figure 33.

Other possibilities of breaking down the two spatial objects considered in the reconstruction method are visualised in Figure 26 to Figure 33, whereby the different internal parts of the spatial objects are marked in different shading colours:

Figure 26 Breakdown of object A in similar[48] objects

Figure 27 Breakdown of object A in congruent objects

Figure 28 Row-wise breakdown of object A in internal parts

Figure 29 Columnwise breakdown of object A

Figure 30 Breakdown of object A in similar / congruent parts

Figure 31 Row-wise breakdown of object A

---

[48]   The meaning of „similar" refers to the way of construction of the different objects, rather than the mathematical meaning.

Figure 32 Breakdown of object B in congruent parts   **Figure 33 Breakdown of object B in different parts**

Concerning the relationship between mental and real spatial abilities, it is assumed that the break-down takes place mentally (visual, cognitive, conceptual) and is translated in words in spatial discourse for communicative purposes, thus this strategy can also be referred to as mental object break-down. Therefore, the language used by the describer represents the externalisation of mental actions concerning the intended break-down of the object. Simultaneously, the describer might also show the break-down of the object by real actions, which makes it easier for the researcher to understand the way students break down the object in their spatial discourse. Nevertheless, the use of this strategy should be identified by the explicit description of internal parts of spatial objects and their underlying construction, which structures the development of spatial discourse between the students in the reconstruction method.

5.1.3 Assembling strategy

A further strategy used to solve the spatial task in the reconstruction method is the description of the *assembling of internal parts* to construct the spatial object. In this strategy, students describe how the parts in which the spatial objects have been broken into should be assembled or put together, which most commonly requires the description of the spatial relation between the spatial parts. This requires the previous use of strategy *object break-down*, whereby the students must describe the break down of the spatial object in several parts. The following transcript excerpt shows the describer's use of the *assembling strategy* in her spatial discourse during the reconstruction method:

Transcript 11:

(Describer 7 has described how to build two internal parts of the spatial object A, which are referred to as *the one with three cubes* and *the one with four cubes*)

Describer 7:   *And then you take this other   Und dann nimmst du dieses An-*
                 one with three cubes.   dere mit drei Würfeln.

Builder 7:	*So the other one, right?*	*Also das Andere, was?*
Describer 7:	*Yes, firstly do it together with the first one of three cubes.*	*Ja, du tust als Erstes zusammen mit dem Ersten mit drei Wür-*
Builder 7:	*How shall...?*	*Wie soll...?*
Describer 7:	*Well the one on the right, yeah?*	*Also bei dem ersten rechts, ne?*
Builder 7:	*Yes, so the one with four.*	*Ja, also bei dem mit vier.*
Describer 7:	*So go to the three cubes again. (...). And at the top, there is a hole and there, you put it in the first one together. So at the first one on the right, do it together to the (...)*	*Also geh nochmal bei den drei Würfeln. (...). Und da oben ist so ein Loch und da steckst mit der ersten zusammen. Also bei dem Ersten rechts mit vier Wür- feln das zusammen. (...)*

In Transcript 11, the Describer 7 describes how one part of the spatial object *the one with three cubes* should be assembled to the second part of the spatial object *the one with four cubes*. The different nominal phrases which act as name-givers to the two spatial parts indicate the previous use of break-down strategy, which is followed by the assembling of both parts. Although the assembling was mere- ly described as 'putting together' in Transcript 11, the student explicitly verbal- ised that both parts must somehow be assembled together for obtaining the de- sired spatial constellation. Whereas, the use of assembling strategy requires break-down strategy, the other way round is not always the case, as Transcript 12, which is a representative of the overall structure of Describer 5's spatial dis- course, shows:

Transcript 12:

(Describer 5 has described how to build one internal parts of the spatial object A, which is referred to as *staircase*)

	(...)	*(...)*
Describer 5:	*And then you do the same staircase again, as you did first...*	*Und dann machst du nochmal dieselbe Treppe, wie du die beim Ersten gemacht hast...*
Builder 5:	*Should I uhm do a new now?*	*Soll ich jetzt äh neu machen?*
Describer 5:	*What [do you mean] doing a new?*	*Wie neu machen?*
Builder 5:	*I mean, again?*	*Also, nochmal?*

Describer 5:	No! Well the same staircase, which you have done, build it again.	Nein! Also diesselbe Treppe, die du gemacht hast, nochmal nachbauen.
Builder 5:	Then, then there are two?	Dann, dann sind das zwei?
Describer 5:	Yes, then there would be two staircases.	Ja, dann wären es zwei Treppen.
Builder 5:	Ok.	Okay.
Describer 5:	Let me know when you are finished with it.	Sag Bescheid, wenn du fertig bist damit.

(End of spatial task A)

In the above transcript excerpt, Transcript 12, the student described how to construct two internal parts of the object, which are referred to as *staircases*. After the builder constructed both internal parts, Describer 5 signalises the end of the description; hence he fails to use the assembling strategy for describing how the two staircases should be assembled together to obtain spatial object A.

5.1.4 Rotation strategy

Another strategy observed during the data analysis of the spatial tasks solving processes is *rotation strategy* which denotes the use of rotation of the object around one of the three axes (vertical, horizontal and frontal) embedded in the describer's spatial discourse. Based on the assumption of transformational equivalence (cf. Finke, 1989; see Section 2.1.2.2), it is assumed that a describer performs the rotation mentally, and gives it a social dimension in space by verbalising it, whilst additionally he or she might perform it using real actions. However, it is important to emphasise that although the describers may perform a lot of actions on the spatial objects, only the rotating actions which have influenced the description process and which are explicitly verbalised in the describer's spatial discourse and addressed to the builder were considered as a rotational strategy.

A complete description of a rotation requires the describer to communicate the degree or type of rotation and the axis around which the three-dimensional object has to be rotated during the reconstruction method. The intention of rotation is to change the spatial position of an internal part of the object during the description process. An example of rotation strategy can be found in the following description of object A in Transcript 13:

Transcript 13:

(Describer 10 is describing to the builder how to construct spatial object A)

Describer 10:	*Then at the top uhm start left too, then three stones.*	*Dann oben äh links auch anfangen, dann drei Steine.*
Builder 10:	*I have.*	*Hab.*
Describer 10:	*Then uh at the top again and then two stones.*	*Dann äh oben wieder und dann zwei Steine.*
Builder 10:	*I have.*	*Hab.*
Describer 10:	*Then you turn it around, so that you... uh... so that it looks away from you, the staircase.*	*Dann drehst du das, dass du... äh... dass es von dir weg zeigt, die Treppe.*
Builder 10:	*Yes.*	*Ja.*
Describer 10:	*Then you build three rows uhm... at the bottom left of the staircase.*	*Dann baust du drei Reihen ähm... von der Treppe unten links.*
	(...)	*(...)*
Builder 10:	*Yes, ok. And when I... now when the staircase looks away from me, so there is uh only a wall then, simply like... like a straight line, right?*	*Ja, okay. Und wenn ich... wenn nun die Treppe von mir weg zeigt, also ist ja dann nur äh nur ne Wand, einfach so... so'n gerader*
Describer 10:	*Yes.*	*Ja.*

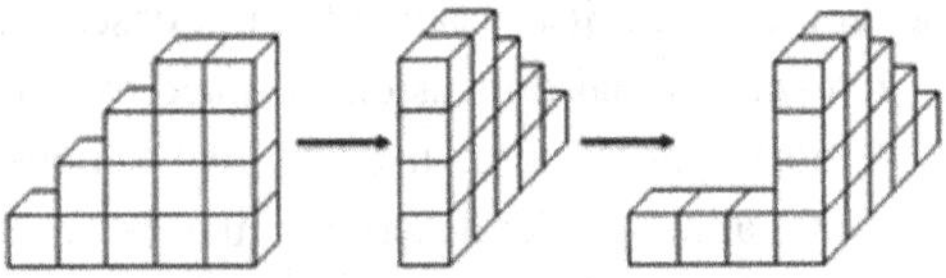

Figure 34 Visualisation of the describer's intended actions during use of rotational strategy on an internal part of the spatial object in Transcript 13

In Transcript 13, the Describer 10 describes how an internal part of the object shall be constructed. The construction of the second internal part was integrated in his description as an 'extension' of the first internal part, and Describer 10 does not describe the spatial relation between both internal parts explicitly by the use of spatial prepositions, but by providing a rotating action on an internal part of the object. The type of rotation was expressed by using a spatial metaphor for its new spatial position: "*it [the internal part of the object] looks away from you, the staircase*" (see Transcript 13). Although this phrase consists of a spatial metaphor based on everyday language (see Section 5.1.1), it entails enough spatial information to indicate the type of rotation and the degree of rotation. The rotation around vertical axis could be conceivably implicitly indicated by the emphasis on the spatial object's structure as *staircase*, whose erected spatial position should be retained. The degree of rotation is expressed by Describer 10's phrase, *"it [the staircase] looks away from you [builder]"*. This phrase reveals that the steps of the staircase should face in an opposite direction from the builder's point of view. For Builder 10 in Transcript 13, this rotation of the internal part of the object is understood and she reassured the describer the spatial-visual perspective of the spatial object in its new spatial position as being *a straight line* after performing the rotational action under consideration of her view of the underlying spatial object.

5.1.5 Cubes controlling strategy

Cubes controlling strategy is one of the controlling strategy observed during the solving of the spatial tasks in the reconstruction method. Kuhl and Beckmann (1985), a self-regulatory or controlling strategy during learning consists of "postdecisional processes that energize and control the maintenance and enactment of intended actions" (Kuhl & Beckmann, 1985, p. 90). Such strategies have the function of controlling the way how tasks are solved and can be considered as metacognitive strategies, since they operate on or regulate the underlying cognitive processes in the solving of the task (cf. Friedrich & Mandl, 2006).

A controlling strategy which was used by describing students for monitoring their task solving process is the questioning about the number of cubes which were used by the builder to construct a spatial object or its internal part, which should be referred to as *cubes controlling strategy* (CSC). The aim of this strategy is to control whether or verify that the actual reconstructed part or object has been built as intended by mapping the spatial object or its internal part to its number of cubes. Hence, this strategy enables a verification of the spatial configuration and reflection on the reconstruction of the object and requires feedback (not in form of yes-or-no answers) from the builder in the reconstruction method when monitoring the current state of spatial construction.

Consider the following transcript, Transcript 14, in which the describer applies the cubes controlling strategy in his spatial discourse in the first spatial task in the reconstruction method:

Transcript 14:

(Describer 10 continues his description from Transcript 13)

	(...)	(...)
Builder 10:	*When I turn it away, there is only the... uh... the fifth row, well the row...*	*Wenn ich die weg dreh, ist doch nur noch die... äh... die fünfte Reihe, also die Reihe...*
Describer 10:	*How many stones have you built to it already? (Counts silently: one, two, ...)*	*Wie viele Steine hast du schon angebaut? (Zählt leise: eins, zwei, ...)*
Builder 10:	*Uh... fourteen.*	*Äh... vierzehn.*
Describer 10:	*That's correct.*	*Ist richtig.*
	(...)	(...)

In Transcript 14, Describer 10 uses the counting controlling strategy by asking the builder how much cubes he has used till a specific point during the task solving process.[49] Following the builder's phrases, *"When I turn it away, there is (...) the fifth row"*, which indicates the previous use of rotational strategy (see Section 5.1.4), Describer 10 used the cubes controlling strategy to verify whether the reconstructed internal part of the spatial object (visualised in on the left hand side of Figure 34) is identical or not with the intended original spatial object. One trigger for the use of this controlling strategy could presumably be the describer's interpretation of the builder's phrase *the fifth row* as consisting of five cubes. However, in this case, the builder was referring to the property of being after each other, rather than the height of the fifth column. Such a different interpretation of the builder's phrase might had been the reason for Describer 10 to use the cubes controlling strategy for controlling the structure of the reconstructed object part.

5.1.6 Structure controlling strategy

Another controlling strategy which was used by the describers to control the description process in the reconstruction method, is *structure controlling strategy* (SCS). In *structure controlling strategy*, a describer demands the builder to describe the structure of the current reconstructed object, which is then used by the describer to compare to the original object, as Transcript 15 shows.

---

[49] The act of questioning characterises the use of the controlling strategies in the reconstruction method. If the describer mentions or states the number of cubes needed to construct the spatial object (e.g., *you need ten cubes*), then this should not be considered as a controlling strategy in the task solving process of the reconstruction method, because it does not indicate any selfmonitoring or self-regulation in the solving process of the task.

Transcript 15:

(Describer 7 is in the middle of her description of spatial object A)

	(...)	(...)
Describer 7:	*Ok. How does it look like? Please tell me that it looks similar to a ladder. Well one which goes down.*	*Okay. Wie sieht das aus? Bitte sag, das sieht ähnlich aus, wie ne Leiter. Also die runter geht.*
Builder 7:	*No.*	*Nein.*
Describer 7:	*One which goes down. Please tell me yes.*	*Die wer runter geht. Bitte sag ja.*
Builder 7:	*It looks like almost... no idea. Wait! (...) I uhh... wait... Well now I have a ladder, where there is a huge staircase, so...*	*Das sieht jetzt ungefähr... keine Ahnung aus. Warte! (...) Ich ahh... warte... Also jetzt hab ich so gesehen eine Leiter, wo eine riesen Treppe ist, also...*
Describer 7:	*Ok.*	*Okay.*
Builder 7:	*Where one, two, three, then I have only two here.*	*Wo eins, zwei, drei, dann hier nur noch zwei habe.*
	(...)	(...)

In the above utterrances, Describer 7 asks the builder, "*how does it [the reconstructed object] look like?*", for comparing the builder's answer with his or her own association of the reconstructed spatial knowledge, which was emphasised by the spatial metaphor with ST-function *ladder*. The builder describes the reconstructed object as, "*a huge staircase*", which seems plausible for the builder presumeably because the structure of a staircase is similar to a ladder (even though the word *huge,* which reveals knowledge about the size of the object, is neglected by the describer). Hence, structure controlling strategy is a strategy which is used by the describer to monitor the task solving process, by involving the builder directly and demanding his or her feedback regarding the structure of the reconstructed spatial object (which does not merely consist of yes-or-no answers or a number of cubes, but of more detailed descriptive feedback).

5.1.7 Discussion of the identified strategies and review of literature

In the above sub-sections, I introduced and described a range of strategies which emerged from the analysis of the data collection in the task solving processes in the reconstruction method. From a researcher's perspective, these strategies describe the describer's actions which are performed to solve the spatial task and represent the flexibility of the individual ways of achieving the desired goal in the reconstruction method of the main study[50]. The first strategy of spatial metaphors entails and represents a large number of such actions, because spatial metaphors as elements of spatial language are useful for communicating spatial knowledge by activating and generating other concepts which evoke other mental images. For instance, spatial metaphors can be useful for identifying internal parts of spatial objects during the break-down strategy. Moreover, the different facets of spatial metaphors have been summarised in the model of spatial metaphors, which visualises the linguistic, the spatial and the conceptual dimensions of spatial metaphors (see Figure 25). The next two identified strategies, breakdown and assembling, are two complementary strategies for a successful task solving process. The separation between both strategies is required due to the fact that subjects may not necessarily have used both strategies during the reconstruction method (see Section 5.1.3). The next strategy, rotation strategy, describes the verbalisation of the object's rotation which is integrated by the describer in his or her spatial discourse. Even though many real rotations of the spatial objects may be performed by the describer, only the verbalised rotating actions which influenced the description are considered as rotation strategies. The last two identified strategies, cubes controlling and structure controlling, are controlling strategies, used by the describers to control the task solving process by demanding feedback from the builders regarding the number of cubes cubes or the structure of the reconstructed object respectively.

The identified strategies provide an overview of the strategic moments during the task solving process in the reconstruction method. As mentioned in the results of the pilot study (see Section 4.1.4), traditional strategy groups such as analytic and holistic are not adequate for a comprehensive differentiation between the strategies used in the reconstruction method. However, Barrat's (1953) differentiation in analytic and holistic can be useful for categorising the identified strategies according to the underlying approach in the task solving process. Holistic and analytic strategies were defined in Section 2.1.5.2 as following:

> Holistic strategies or *spatial manipulation* (cf. Burin, Delgado & Prieto, 2000), are strategies in which students mentally transform or manipulate

---

[50]   Due to the important role of the spatial objects' design for evoking different strategies, further strategies or a different categorisation might be identified if further different structures of spatial objects are considered.

the whole object or its parts or move around the object in order to solve the spatial task (cf. Barrat, 1953). In analytic strategies, also known as *feature comparison* (cf. Burin et al., 2000), students compare details of the orgininal object and then decide whether the original and the comparing object are identical (cf. Barrat, 1953). (Section 2.1.5.2, pp. 31)

Due to the fact that the identified strategies are realised using spatial language, the focus for their theoretical categorisation should be on the verbalisation of holistic or analytic strategies.[51] As pointed out in earlier research (e.g., Plath, 2014), spatial tasks demanding logical reasoning and verbalisations tend to require students to use more analytic than holistic strategies. In contrast, the verbalisation of holistic strategies seems to be more cognitively demanding, because it requires the students to verbalise transformation, manipulation or movement of the spatial objects (cf. Schwank, 2003). In verbalising holistic strategies, students are required to verbalise mental manipulating or transforming actions, which, in contrast to properties or characteristics of spatial objects, tend to be activated more in the image system, rather than as specific terms in the verbal system in the student's cognitive system for information processing (see Section 2.1.2.3).

In the strategy of spatial metaphors, the describers use spatial metaphors to point out, highlight and communicate particular spatial characteristics or properties of the spatial objects in their discourse. The verbalisation of this transfer of properties from the source to the target domain is based on conceptual thinking in metaphors, which are applied on a spatial context. Hence, spatial metaphors would be more likely verbal-analytic rather than holistic.

The complementary strategies, object-breakdown and assembling strategies, can be assigned to the group of holistic strategies, since the describer breaks down the object in separate internal parts and manipulates them wholly, followed by a description of a synthesis of the parts in order to obtain the whole spatial object. Although object-breakdown strategy can be interpreted as a form of analysis, since the object is broken down in different parts and not considered wholly, Barrat's (1950) definitions of analytic and holistic are considered as a fundamen-

---

[51] One must consider the fact that Barratt's (1953) definition of analytic and holistic strategies is based on the comparison between two objects in spatial tasks. Whilst this is not necessarily the main focus in the underlying data analysis of the reconstruction method in this present study, an analysis of the comparison between the original and the reconstructed object could also be possible in the reconstruction method.

tal basis for classifying between analytic and holistic strategies (see Section 2.1.5.2). Hence, the describer is required to verbalise the transformation and manipulation of the spatial object, which consists of the externalisation of mental actions concerning break-down and assembling of objects by using spatial language. In fact, in contrast to spatial metaphors, the verbalisation of the break-down and assembling seems to be more cognitively demanding, because it structures the whole description process and requires the students to verbalise actions manipulating the parts of the object, rather than just describing properties of the spatial objects. Similarly, the strategy of rotation can also be considered as providing a holistic approach to solving the spatial task in the reconstruction method. Describers using the rotation strategy are required to verbalise actions which transform the spatial position of spatial parts, hence it belongs to the group of holistic strategies. In contrast, in the controlling strategies, cubes and structure controlling strategies, the describer tends to evoke strategic moments which conceivably focus more on the feature comparison between the original and the reconstructed object, by referring to the structure or to the number of cubes. Hence, from this perspective, both controlling strategies can be regarded as being of analytic nature. The following table shows a summary of the categorisation of the identified strategies according to Barrat's (1953) holistic and analytic strategies:

Holistic Strategies	Analytic Strategies
Object Breakdown	Spatial Metaphors
Assembling	Cubes Controlling
Rotation	Structure Controlling

Table 20   Categorisation of identified strategies in holistic and analytic dichotomy

A consideration of other dichotomy of strategies, such as *three-dimensional* vs. *two-dimensional* strategies (cf. Gittler 1984, see Section 2.1.5.3) is less adequate for grouping the identified strategies. Two-dimensional and three-dimensional strategies were described as following:

> In *three-dimensional* strategies, students solve the spatial task by developing three-dimensional mental models, which are used, transformed, moved etc… for solving the spatial task (cf. Gittler, 1984). In contrast, students using *two-dimensional* strategies produce two dimensional images of the three dimensional spatial object to facilitate the solving of the spatial task (cf. Gittler, 1984), hence neglecting the third dimension in space. (Section 2.1.5.3, p. 33)

154

Due to the presence of three dimensional models in the spatial tasks of the reconstruction method, it is assumed that describers presumeably develop three-dimensional mental models of the spatial object, on which they apply strategies such as break-down and assembling. However, an indication of a two dimensional approach may be found in the use of spatial language, such as the choice of spatial metaphors, for describing three dimensional models.[52] For instance, consider the spatial metaphor *line*[53] in the following excerpt from Transcript 13 (in Section 4.1.4):

Builder 10:   *Yes, ok. And when I... now when the staircase looks away from me, so there is uh only a wall then, simply like... like a straight line, right?*   *Ja, okay. Und wenn ich... wenn nun die Treppe von mir weg zeigt, also ist ja dann nur äh nur ne Wand, einfach so... so'n gerader Strich, oder?*

In the above utterrances, Builder 10 uses a term of a two-dimensional nature, in order to describe a three-dimensional model from a particular perspective (after using rotational strategy). Other concepts originating from the two-dimensional world include *triangle* (see Transcript 5), *quadrilateral* and other name of figures. Therefore, the categorisation of two-dimensional and three-dimensional thinking could be conceivably applicable to the category of spatial metaphors. However, due to the increase in complexity of the model of spatial metaphors (in Section 5.1.1.4), the differentiation between spatial metaphors of two-dimensional and three-dimensional nature as the fourth dimension should be neglected in the present model for spatial metaphors. For this reason, the holistic and the analytic dichotomy was considered as the most adequate for grouping the strategies identified in the data analysis process of the main study (see Table 20).

The identified strategies represent a wide range of strategies applied by students when describing spatial objects, however, further strategies can exist which

---

[52]   To a certain extent, the strategies of breakdown and assembling can be also considered as "two-dimensionalising" and "three-dimensionalising" processes in spatial discourse respectively.

[53]   In the German language, the term *Strich* is rather used in everyday language. The equivalent terms in mathematics language of secondary school students would rather be the word *Strecke* (as a line with starting and ending point) or sometimes even *Gerade* (the actual mathematical meaning is of a line with no starting or ending point), which is oftenly used among students to refer to *Strecke*.

might not have been identified in the describing students' spatial discourse in the sampling group of this present study. An example of such a strategy can be the codification of directions along the three dimensions for representing a reference frame for describing the construction of the spatial objects (e.g., 1 for right, 2 for left, 3 for behind, 4 for in front of etc. used by a builder in a spatial task solving process; see Transcript 16 in Section 5.2.2). Nevertheless, due to the relatively wide range of sampling considered in this present study the above identified strategies should be deemed as characteristic for fifth-grade students solving spatial-verbal tasks in the reconstruction method.

5.2 Description of identified obstacles

After having described the strategies used by students during the task solving process in the reconstruction method, obstacles encountered by students during this process are discussed, as indicated in the next research question:

(R2)   Which obstacles do students encounter during the description of spatial configurations in spatial-verbal tasks?

The term *obstacles* or *barriers* denotes the moments or choices which hinder rather than support the task solving process concerning the desired output in the solving of spatial tasks in the reconstruction method. The identification and understanding of obstacles can act as diagnostic tools for understanding the students' spatial concepts and, similar to strategies, they can be detected by an analysis of spatial language in discourse. Obstacles can emerge in communicative processes between students in the reconstruction method, given their different experiences and background knowledge which lead to the generation of different interpretations and hence different reconstruction of knowledge from a constructivist approach. Before pointing out the obstacles encountered by the students describing spatial objects in the reconstruction method, a model for characterising the different cognitive processes in which obstacles can emerge in the solving process of the spatial task in the reconstruction method is illustrated in Figure 35.

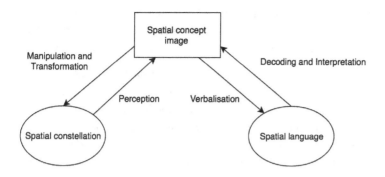

Figure 35 A model of the spatial-visual processes for identification of obstacles in spatial tasks of the reconstruction method

In general, spatial knowledge and thinking is internally represented within the spatial concept images in the individual's cognitive system, which are partially externalised, and hence observable either by the individual's verbalisation via spatial language or by the individual's real actions on spatial objects in the re-construction method. The process of visual perception of spatial objects de-scribes the student's awareness of particular spatial knowledge which he or she activates when dealing with the spatial object (cf. Lurija, 1992). Activated men-tal concept images and mental actions on spatial objects trigger the manipulation and transformation of spatial objects, which in turn might lead to the perception of new spatial knowledge. The externalisation of spatial knowledge via spatial language is another way of how spatial knowledge can obtain a social dimen-sion. The verbalisation of spatial thinking and knowledge requires its interpreta-tion, which demands the activation of mental spatial images and knowledge for its understanding. Hence, the perception, manipulation, and transformation of spatial objects, as well as the verbalisation, encoding, and interpretation of spa-tial language, demand the activation of spatial concept images and contribute to the further development of spatial concepts in the individual. Moreover, verbali-sation and manipulation or transformation of the spatial object enable the re-searcher to observe, understand, and create assumptions about the students' spa-tial concept images and their underlying meanings in spatial discourse (i.e. per-ception and interpretation processes) (see Figure 35). Three major moments in which obstacles emerge have been identified in task solving processes of the re-construction method: (1) during the perception of a particular spatial object, and / or (2) during the process of verbalisation of a particular spatial concept image, and / or (3) during the encoding and interpreting of communicated spatial knowledge in order to reconstruct the spatial concept image, and / or (4) during the manipulation and transformation of the spatial objects (see Figure 35).

In the following sub-sections, exemplary students' obstacles are described in detail by referring to different case studies for a deeper understanding of obstacles which emerged in spatial discourse in the reconstruction method. The wide range of identified obstacles, i.e. *spatial metaphors* (as obstacles), *describing spatial relations*, *verbalising rotations*, *dimension reduction*, and *spatial disorientation*, should not be considered as fixed or complete, but rather as an explorative approach for identifying and classifying main obstacles in the solving process within the reconstruction method.

5.2.1 Spatial metaphors as obstacles

As discussed in Section 5.1.1, spatial metaphors are used by students to solve spatial tasks in the reconstruction method. Whilst spatial metaphors can be used to bridge the barriers during the solving of spatial tasks (see Section 5.1.1), they can also act as obstacles themselves. The advantage of spatial metaphors in spatial discourse is that they create an image in the listener's mind which can be generated, communicated, evoked and activated by language easily. However, only the generation of mental images and their communication is not sufficient for an extensive spatial description, especially when the spatial metaphor used can evoke another mental image in the listener's mind rather than the describer's intended image. Therefore, spatial metaphors can also act as obstacles during the process of decoding interpretation and of the underlying spatial knowledge or properties entailed in the spatial metaphor. Consider the following transcript excerpt, in which the student describes the construction of spatial object A using the spatial metaphor *wall*:

Transcript 16:

(the student gets to take a glimpse at spatial object A and attempts to improve her description, by starting to describe how the two congruent internal parts should be constructed to each other)

| Describer 8: | *So like staircases which look like back-to-back or so.* | *So wie die Treppen so wie Rücken einander aussehen oder so.* |

Builder 8:	That means that I must do it now... on the staircase, on it. [see Figure 36 for builder's reconstructed object at this time]	Das heißt ich muss die jetzt... auf die Treppe noch draufmachen. [siehe Abbildung 36 für das nachgebaute Objekt zu diesem Zeit-
Describer 8:	No.	Nein.
Builder 8:	Right?	Oder?
Describer 8:	Next to it completely well you look... well you can see...	Daneben ganz also du guckst... du siehst ja...
Builder 8:	This means that the last foursome staircase, over there are still two at it.	Das heißt, die letzte Vierertreppe, da sind zwei dann noch dran.
Describer 8:	No! Uh... wait... you must... you do it in a way that the foursome staircase looks like a wall, so that then well uhm...	Nein! Äh... warte... du musst... du mach das mal so als ob die Vierertreppe wie eine Wand aussieht, dass das dann so ähm...
Builder 8:	This means that then I don't have a cube anymore to go up on?	Das heißt, dass ich dann keinen Würfel mehr hab zum hochgehen?
Describer 8:	From your sight, it should be left or so, so that the foursome staircase uhm...	Von dir muss es eigentlich links oder so sein, dass die Vierertreppe ähm...
Builder 8:	This means that I must actually have four, eight, twelve cubes together... a kind of... chocolate bar then... [see Figure 37 for the builder's reconstructed object at this time]	Das heißt, ich muss eigentlich vier, acht, zwölf Würfel aneinander haben... wie so ne Art... Tafel Schokolade dann... [siehe Abbildung 37 für das nachgebaute Object zu diesem Zeitpunkt]
Describer 8:	No!	Nein!
	(...)	(...)

Figure 36 Visualisation of Builder 8's reconstructed object in Transcript 16 (Part One)

Figure 37 Visualisation of Builder 8's reconstructed object in Transcript 16 (Part Two)

In the above transcript excerpt, Transcript 16, Describer 8 describes the construction of the internal parts, one of them is referred to as *foursome stairs*, and uses the spatial metaphor *wall* to describe the spatial relation between the two internal parts, after several unsuccessful attempts (first by using the spatial metaphor(s) *back-to-back stairs* at the beginning of Transcript 16). In order to describe this spatial relation, Describer 8 suggests that the foursome stairs should look like *wall*, however, the spatial metaphor *wall* was associated to another meaning and hence generated another image from the builder's point of view. For the builder, a *wall* resembles a cuboid, which he indicates in the phrase, *"This means that I don't have a cube anymore to go up on?"*. Hence, he suggested the expansion of one internal part of the object, *the foursome stairs*, to a cuboid which he described by using the spatial metaphor with ST-function, *bar of chocolate* (see Figure 37 for the builder's interpretation of the metaphor *wall* in the reconstruction of the spatial object). Upon the mentioning of this new metaphor, the describer discardes the spatial metaphor *wall* and attempts another way to describe the spatial relation between both internal parts. This episode in Transcript 16 shows that whereas spatial metaphors can be useful to communicate spatial knowledge in a conventional way, spatial metaphors are not unambiguous, since they enable a high number of interpretations, which are reconstructed depending on the concerning subject's experience and knowledge. With reference to the model illustrated in Figure 35, decoding and interpretations of spatial metaphors can lead to obstacles by evoking undesired spatial concept images, which once activated influence the manipulation and transformation of the spatial object by the builder. Therefore, a high use of spatial metaphors might be hindering instead of supporting for communicating spatial knowledge.

5.2.2 Describing spatial relations

The verbalisation of spatial relations is an important process during the solving process in the reconstruction method. As explained in previous sections (see Section 2.1.2.1), the cognitive processes of identifying and verbalising spatial relations requires at least two spatial objects in space and the perception of their relation to each other, whereby the spatial objects in the reconstruction method can consist of single cubes or internal parts of the designed objects (see Section 4.2.2.3). However, for various describing students, the explicit verbalisation of spatial relations was not emphasised enough in their spatial discourse, as the following transcript excerpt, Transcript 17, shows:

Transcript 17:

(Describer 3 starts describing spatial object A)

Describer 3:	*After a cube, take two cubes. Then stick two cubes together firstly and then another cube next to them.*	*Nach einem Würfel nimm zwei Würfel. Also kleb zwei Würfel erstmal zusammen und danach noch ein Würfel neben die.*
Builder 3:	*How? At which side? Top, bottom, left, right?*	*Wie? An welche Seite? Oben, unten, links, rechts?*
Describer 3:	*Uhm... stick it!*	*Ähm... Kleben!*
Builder 3:	*Yes, top, bottom, left, right?*	*Ja, oben, unten, links,*
Describer 3:	*Yes, how shall I explain it to you?*	*Ja, wie soll ich dir das erklären?*
Builder 3:	*Left-right tactic!*	*Links-rechts Taktik!*
Describer 3:	*How shall I explain it to you, do you know?*	*Wie soll ich dir das erklären, weißt du?*
Builder 3:	*One is left and two is right.*	*Eins ist links und zwei ist rechts.*
Describer 3:	*Stick it simply!*	*Kleb einfach das eine!*
Builder 3:	*Yes, I already have it so two.*	*Ja, ich habe das schon so zweier.*
Describer 3:	*A cube next to that... uh... how shall I explain it? I don't know how to explain it to you. (...) It is a ladder!*	*Würfel neben das... äh... wie soll ich dir das erklären? Ich weiß nicht, wie ich dir das erklären soll. (...) Das ist ne Leiter!*
	(...)	*(...)*

In Transcript 17, Describer 3, starts by instructing the builder how to start constructing spatial object A, by referring to two cubes and one cube, which should be assembled together. Upon the builder asking for a more precise spatial relation between the two-cubed part and the cube, the describer presumeably failed to realise the ambiguity regarding the spatial relation expressed using the spatial preposition *next to* in her spatial discourse. Although the builder reveals once again this ambiguity in her utterance, "*Yes, top, bottom, left, right?*" (even by trying to introduce codes for left or right orientation in "*One if left, and two is right*"). Describer 3 does not take the possible different relations into account, but rather focussed on finding a spatial metaphor, in this case, *ladder*, which describes the whole spatial configuration.

However, sometimes students failed to describe the spatial relation at all in their spatial discourse. Consider, for instance, the student utterrances in the episode illustrated in Transcript 12 in Section 5.1.3. In this excerpt, the student breaks down the spatial object into two internal parts. When the builder finishes constructing both parts, the describer fails to describe the spatial relation among each other, thus as predicted the builder ends up reconstructing two separate spatial objects, two *staircases* (see Transcript 12 in Section 5.1.3). The description of how to connect the two object parts, in which the spatial object has been mentally broken into, is an important demand for the use of the assembling strategy, which follows the strategy of break-down. However, a successful use of this strategy requires a rich description of the spatial relation, not merely the fact that the object parts should be somehow connected (which is sufficient for identification of use of assembling strategy; see Section 5.1.3). Consider, for instance, the following transcript excerpt, which is a continuation of Transcript 11, whereby the describer instructs the builder to assemble the two internal parts of spatial object A together (in Section 5.1.3):

Transcript 18:

(Describer 7 disassembles the parts of spatial object A and does the movement to reconstruct the whole object again)

	(...)	*(...)*
Describer 7:	*And there at the top, there is a hole and there you put it with the first one together. So at the first one with four cubes, that together.*	*Und da oben ist noch ein Loch und da steckst es mit der Ersten zusammen. Also bei dem Ersten rechts mit vier Würfeln das zusammen.*
Builder 7:	*How shall I put it?*	*Wie soll ich das hinstellen?*
Describer 7:	*Straight, well uhm straight ahead, simply put it in*	*Gerade, also ähm geradeaus, einfach normal reintun.*
Builder 7:	*Like this?*	*So?*
Describer 7:	*How? Well you take the first one and put it in the first one [of the other part] together.*	*Wie? Also du nimmst das Erste und tust du bei dem Ersten rechts rein zusammen.*

| Builder 7: | At the… so I take the small one and do it with the big one together? | Bei dem… Also ich nehme den Kleinen und tue es mit dem Großen zusammen? |
| Describer 7: | Yes, but only the one at the right, there where there are four cubes. (…). Ok that was it. | Ja, aber nur dem ersten rechts, da wo vier Würfel sind. (…). Okay das war es. |

In Transcript 18, Describer 7 attempts to describe the spatial relation between both objects, which had already been constructed by the builder and were referred to as *the small one* and *the big one*. The description was characterised by the ongoing actions which were verbalised by Describer 7, such as, *"put it with the first one together"*, whereby Describer 7 does not seem to include the static spatial relation between the objects. The describer's utterrances were characterised by high spatial ambiguity, which she does not seem to fully recognise, even though she gives hints that the spatial relation is on the right-hand side of the *first* part. Nevertheless, this information was not enough for the builder, who iterated, *"How shall I put it?"*, since there was more than one possibility of connecting both internal parts. The describer's reply to the builder's question, *"straight (…) straight ahead (…) simply put it in normal"*, was not sufficient for the description of the orthogonal spatial relation between both internal parts, especially since *normal* in the last phrase did not carry any spatial meaning. The above ambiguity regarding the spatial relation between the two internal parts led to misunderstandings which were noticed in the builder's rebuilt object (see Figure 38).

Figure 38 Builder's reconstructed spatial object in Transcript 18

The above reference to excerpts from student's spatial discourses shows how possible obstacles can arise when spatial relations are either not or rather partially described in spatial context, especially when a reference system is missing in the spatial discourse. As discussed, spatial relations can be expressed using spatial prepositions or spatial metaphors (see Section 5.1.1) or both. However, a description of a spatial relation also requires a high level of spatial awareness and the understanding of possible ambiguity in space, especially when two parts need to be connected together, whereby the spatial position of the separate internal parts must also be considered. Students need to understand that when connecting two parts to each other, there is more than one possibility of doing this and not all of them lead to the construction of equivalent spatial objects. Hence, this issue should be addressed and developed in spatial discourse for reducing ambiguity and increasing students' spatial awareness in solving spatial-verbal tasks. The degree of spatial (non-)ambiguity in spatial discourse is expected to be identifiable or accounted for by an analysis of linguistic elements of students' spatial language, which are addressed in-depth in structural approaches of spatial language in Section 5.4.

5.2.3 Verbalising rotation

Another obstacle identified concerned the describer's verbalisation of a rotating action in the reconstruction method. As described in Section 5.1.4, describers making use of this strategy must verbalise the rotation in their spatial discourse, however, some describers could not verbalise the intended rotation during the reconstruction method, as it can be noticed in Transcript 19 (wherein Describer 9 is describing spatial object B):

Transcript 19:

(Describer 9 instructs how to obtain the two identical parts of spatial object B, which he breaks down according to Figure 34).

(...)	*(...)*

Describer 9: *Do this tower apart, in the middle, half of it, so that you have a tenth on one side and the same on the other side. And then you put them longish to the top in front of you, but so that, how shall I say it... (shows movements with his hands parallel and orthogonal to his front) It's difficult...* | *Mach diesen Turm auseinander, in der Mitte, die Hälfte, so dass du auf einer Seite ein Zehner hast und auf der anderen Seite auch. Und dann stellst du die länglich nach oben so hin vor dir, aber so, dass, wie soll ich das sagen... (bewegt seine Hände parallel und orthogonal zu seiner Vorderseite) Schwierig...*

(...)	*(...)*

Describer 9:	*Now you still have the ten towers, the two?*	*Jetzt hast du noch die Zehnertürme, die zwei?*
Builder 9:	*Yes.*	*Ja.*
Describer 9:	*Now you put in such a way that...* (repeats the same movement) *here that they... look to the top first. Do you have it?*	*Jetzt stellst du die so hin, dass sie...* (wiederholt die gleiche Bewegung) *hier dass sie... nach oben gucken einmal. Hast du?*
Builder 9:	*Yes.*	*Ja.*
Describer 9:	*And then you do the twosome inbetween, the third from the bottom.*	*Und dann machst du dazwischen die Zweier das dritte unten.*
	(...)	*(...)*

Figure 39 The describer's intended rotation of two internal parts of spatial object B in Transcript 18 (left: actual position; right: intended position)

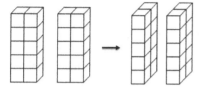

Figure 40 Gestures performed on the spatial object supporting the description of the intended spatial actions in Transcript 19

In Transcript 19, Describer 9 first describes the construction of the two internal parts of spatial object B, each of which consisting of *two foursome towers together at the sides,* which, as the describer has predicted, was reconstructed in the spatial position visualised on the left side of Figure 39. To describe how the spatial parts should be positioned or placed in space, he starts applying words such as *longish* and *looking to the top* (intended meaning is erected). However, he reaches his linguistic limitation regarding the verbalisation of a particular characteristic, which required the analysis of the used gestures. When the describer could not verbalise the underlying phenomenon, he switches to gestures and holds his hands along the spatial object as indicated in Figure 40, in order to visualise that both internal parts should be rotated by 90 degress around the vertical axis. However, Describer 9 does not seem to have a strong command of spatial linguistic repertoire and after the pause attempts to verbalise this spatial characteristic or action using merely the metaphor *looking up* in his spatial discourse (see Transcript 19). The episode in Transcript 19 shows the importance of language for expressing spatial phenomena, which influences the solving of this spatial-verbal task. Even though, the describer shows relatively high spatial awareness (which can additionally be accounted for by the describer's high results in the reference tests of spatial abilities), the linguistic limitation hindered him from communicating the intended spatial knowledge required for successfully solving the task. Hence, this episode indicates the students' difficulty of verbalising the holisitic strategy of rotation during the solving of the spatial verbal-tasks in the reconstruction method, which concurs well with Schwank's (2003) findings about the increased obstacles when students verbalise functional or holistic thinking.

The next episode illustrates how describers might perform rotating actions and not attempt to verbalise them, presumeably due to lack of spatial awareness, as the following transcript excerpt illustrates.

Transcript 20:

(Describer 13 gets a glimpse of the builder's reconstructed object, which is visualised below, and attempts to change his description of spatial object A).

	(...)	(...)
Describer 13:	*Firstly, four stones in a row.*	*Erstmal vier Steine in einer Reihe.*
Builder 13:	*Yes.*	*Ja.*
Describer 13:	*Three on it in the right... and the left one must remain empty, the left stone.*	*Darauf drei in die rechte... und das linke muss frei bleiben, der linke Stein.*

Builder 13:	*Yes.*	*Ja.*
Describer 13:	*Two on it, and then the left one remains empty again.*	*Darauf zwei, und der linke bleibt wieder frei.*
Builder 13:	*Yes.*	*Ja.*
Describer 13:	*And then one stone, and the left one remains empty again. (rotates the spatial object 90 degrees around the vertical axis)*	*Und dann ein Stein, und das linke bleibt wieder frei. (dreht das Objekt um 90 Grad um die vertikale Achse)*
Builder 13:	*Yes.*	*Ja.*
Describer 13:	*Now you build at... at the top most stone one at it... at the side.*	*Jetzt baust du an... an dem obersten Stein eins dran... an die Seite.*
Builder 13:	*Left or right?*	*Links oder rechts?*
Describer 13:	*Left. (...) Under it under the left stone there are two again.*	*Links. (...) Darunter unter dem linken Stein kommen wieder zwei.*
Builder 13:	*Yes.*	*Ja.*
Describer 13:	*Under it there are three under the one which you have just built.*	*Darunter kommen drei unter die du grad gebaut hast.*
	(...)	*(...)*

In Transcript 20, Describer 13 describes the construction of the spatial object by breaking it down into rows, although the information that the parts should be horizontal is explicitly emphasised only once in his spatial discourse.[54] The transition from one internal part to another in his description, which he does not explicitly mention, required a change of dimensionality description. However, Describer 13 rotates the original spatial object (see Figure 41) and does not verbalise these actions, which are essential for a successful solving of the task. Instead he merely instructs the builder to build a cube at the left-hand side of the top most cube, while he continues with the description.

---

[54]   In fact, this can also be seen as an obstacle which emerges from the reluctance of emphasising the spatial position of the internal parts. Whereas the describer includes the spatial relation by using spatial prepositions, for example, *underneath* or *under it,* he fails to remphasise the spatial position of the internal parts to reduce spatial ambiguity.

Figure 41 Describer's rotating actions during the description of spatial object A in Transcript 20 (left: initial position, right: final position)

The episode in Transcript 20 shows the importance of rotating actions performed on spatial objects and their influence on the descriptive process. The verbalisation of such spatial manipulations requires their perception and a high level of spatial awareness, which reflects the interplay of spatial abilities and language in these spatial-tasks. As the next episode in Transcript 21 shows, some describers explicitly verbalising rotating actions might also face obstacles in finding the right words to express the underlying phenomenon.

Transcript 21:

(Describer 9 takes a glimpse at the reconstructed object and attempts to change the description of spatial object A).

	(...)	*(...)*
Describer 9:	*Yes... So where which I was... what I said at last, you must turn it around so that it becomes a staircase.*	*Ja... Also wo was ich am... als letztes gesagt hab, das musst du umdrehen so dass das wie eine Treppe wird.*
Builder 9:	*On which side?... How turn it now...*	*Auf welche Seite?... Wie jetzt umdrehen...*
Describer 9:	*You have... you have it uhm done it so that it doesn't look like a staircase at the side, what I have explained to you lastly.*	*Du hast... du hast das doch ähm so gemacht dass das nicht wie eine Treppe aussieht an einer Seite was ich als letztes erklärt hab.*

| Builder 9: | *Indeed, I have one, two, three, four high and... then at the other side one, two, three high.* | *Doch, hab ich eins, zwei, drei, vier hoch und... dann an der anderen Seite eins, zwei, drei hoch.* |
| | *(...)* | *(...)* |

In the above excerpt, Describer 9 attempts to describe the movement or rotation of one internal part of the object, whose spatial relation to the other part has not been reconstructed successfully. The description of the rotation is incomplete, since the reference to the axis around which the object should be rotated is missing. It can be argued that the use of the spatial metaphor with primarily S-function *staircase* in the utterances, "*turn it around so that it becomes a staircase*", could compensate for the missing spatial information. However, the describer's reliance on the use of spatial metaphors to describe this rotating action is not successful and the use of a spatial metaphor for determining the rotation acts as an obstacle to describe the actual intended spatial constellation. In fact, Builder 9 quickly replies that he has in fact two pairs of stairs and reassured the describer by counting the height of the steps.

The above analyses show how rotation of spatial objects plays an important role in the spatial task solving of the reconstruction method. However, as the above episodes illustrated, obstacles emerge once spatial objects are rotated during the description, but these actions cannot be verbalised properly or if they are not verbalised at all due to linguistic limitations or lack of spatial awareness.

5.2.4 Dimension reduction

Another phenomenon which can be characterised as an obstacle in spatial discourse is the negligence of important spatial characteristics, such as those which characterise the third dimension of the spatial object in spatial discourse. As already described in the design of spatial object in the main study (see Section 4.2.2.4), dimensionality was an important characteristic for the design of spatial objects, whereby the students are required to describe along all three dimensions. However, some students, for instance, the Describer 9 in Transcript 22, seem to neglect important spatial characteristics of spatial objects which led to the phenomenon which shall be referred to as dimension reduction.

Transcript 22:

(Describer 9 gets a glimpse of the builder's reconstructed object, which is visualised below, and attempts to change his description of spatial object A).

	(...)	(...)
Describer 9:	*The foursome tower, not to the top, but from the side. (...). And then you do the threesome at the side at it. Have you done it?*	*Den Viererturm nicht nach oben sondern von der Seite. (...). Und dann machst du den Dreier an der Seite dran. Hast du?*
Builder 9:	*So like the foursome, right?*	*So wie die Vierer, ja?*
Describer 9:	*Yes, have you got it?*	*Ja. Hast du?*
Builder 9:	*I have.*	*Habe ich.*
Describer 9:	*(...) Do two double and two single.*	*(...) Mach zwei Zweier und zwei Einer.*
Builder 9:	*Ok, these also next to each other?*	*Okay. Die auch nebeneinander?*
Describer 9:	*No, yes! (...)*	*Nein, doch (...)*
Builder 9:	*I have the foursome towers now, and then the triples*	*Ich habe jetzt die Vierertürme und dann noch die Dreier*
Describer 9:	*And then after the triples, the doubles.*	*Und dann nach den Dreiern, die Zweier.*
Builder 9:	*Yes, I have done it.*	*Ja, habe ich gemacht.*
Describer 9:	*And then after the double, the single.*	*Und dann nach den Zweier ein Einer.*
Builder 9:	*So left and right build the double at it and then the singles, right? (...)*	*Also links und rechts das Zweier dran bauen und dann die Einser, ja? (...)*

In Transcript 22, Describer 9 breaks down the spatial object into single columns, and uses the nominal phrase, "*next to it*", to describe the spatial relation between the columns. Hence, the student described spatial object A as connecting different columns either on the right or left-hand side of the existing ones, as it had been confirmed by the builder in the last utterance of the transcript excerpt,

Transcript 22.[55] Therefore, the describer two dimensionalised the spatial object by neglecting an important property required for a 'three-dimensional' spatial discourse. As a matter of fact, the reconstrcted object at the end of the spatial task A of Describer 9 and Builder 9 can also be considered as two dimensional under negligence of its (see Figure 38). There might be different reasons for the emergence of this phenomenon. Whilst some students deemed an intentional neglection of some spatial properties as a facilitation of the descriptive process, other students might not have perceived the neglected spatial property of the object space.

5.2.5 Space disorientation

Spatial orientation is an important spatial abilty which was described in the theoretical framework as the ability to think about those spatial relations in which the body orientation of the observer is an essential part of the problem (cf. Thurstone, 1950 in Section 2.1.3.1). The ability to orientate oneself in space is very important not only for describing spatial objects in space, but also for decoding and interpreting the given instructions and reconstruct the intended spatial object. The student's back-to-back spatial position in the reconstruction method has been an additional obstacle for some students to orientate themselves in relation to others in space, as can be observed in the following excerpt:

Transcript 23:
(Describer 8 is in the middle of her description of spatial object A).

	(...)	*(...)*
Describer 8:	*So and then you must do one on it, the right one.*	*So und dann musst du eine da drüber das rechte.*
Builder 8:	*Do you mean my right or your right?*	*Meinst du von dir aus rechts oder von mir aus*
Describer 8:	*What do you mean now?*	*Wie meinst du das jetzt?*
Builder 8:	*We are mirror-inverted.*	*Wir sind ja spie-*
Describer 8:	*Yes, for you it is then... left or so or right, I have no clue... uh?... how are you holding it at the moment?*	*Ja, von dir aus ist es dann... links oder so oder rechts keine Ahnung... äh?... wie hältst du es*
	(...)	*(...)*

In the above dialogue between Describer 8 and Builder 8, the students fail to recognise the difference between back-to-back and back-to-front position in

---

[55]   A visualisation of the builder's reconstructed object can be seen in Figure 38.

space, which resulted in uncertainty about the right and left orientation in spatial discourse. Whilst this misunderstanding is triggered by the builder's assuring statement that they were sitting in a mirror-inverted position, the describer seems not to have understood the meaning of mirror-inverted, which leads to confusion from her perspective. Hence, this obstacle of spatial disorientation, whereby students had difficulties in orientating themselves in space (in this case, differentiating between left and right discrimination in a particular spatial setting), was another obstacle which students could encounter in solving of spatial tasks in the reconstruction method and which was triggered by the required spatial position of the same data collection method.

5.2.6 Discussion of the identified obstacles

In the above sub-sections in Section 5.2, I have attempted to provide a wide range of possible obstacles which arose when students describe spatial objects during the spatial task solving process in the reconstruction method. Whereas their identification and explanation can be highly influenced by the design of the spatial object, on a meta-level, the obstacles represent conflicts which might arise during different processes in students' spatial learning. Obtacles can arise during the verbalisation of spatial knowledge, such as of spatial relations (see Section 5.2.2) or spatial actions, in particular, of holistic nature, such as of rotation of spatial objects (see Section 5.2.3). Other obstacles are related to the understanding of spatial language, such as the interpretation of spatial metaphors in spatial discourse. In particular, spatial metaphors can activate different spatial mental images depending on the student's experience and pre-knowledge (see Section 5.2.1). Further obstacles can arise from a different perception of the spatial object, whereby students perceive some spatial characteristics and neglect others, which are not considered crucial in spatial discourse as the former ones. Such an example of this phenomenon, dimension reduction, was discussed in Section 5.2.4. Another example, which might be considered as a misperception of space or spatial concepts was provided in Section 5.2.5, whereby the issue of the insecure grasp of spatial orientation led to the confusion between back-to-back and back-to-front position. Whereas other possible obstacles were noticed, for example, the use of the term *quadrilateral* for different three-dimensional models (which could be interpreted as a form of dimension reduction), the above findings provide a first approach for categorising obstacles in solving spatial tasks in spatial discourse.

172

Overall, the above identified obstacles show that students' spatial language is not fully developed and requires further support and fostering in mathematics classroom. Especially the obstacles related to the verbalisation of spatial knowledge and actions show that language and content learning is necessary in geometry learning and teaching. This does not necessarily imply that students merely learn new vocabulary, but that they reflect upon their use of language for expressing geometrical-spatial concepts and become aware of the necessity of learning mathematical language. The fact that spatial metaphors, which were identified as possible strategies among students describing spatial objects (see Section 5.1.1.1), can also act as barriers show the importance for students to develop their mathematical language in spatial discourse, which is characterised by its exactness and a low level of ambiguity regarding its meaning in discourse.

Moreover, the investigation of obstacles for a deeper understanding of conflicts arising in solving verbal spatial tasks confirms the close interlinking between language and spatial ability, whereby spatial language should not merely be considered as the externalisation of spatial thinking by its linguistic structures, but perceived spatial knowledge or spatial thinking can also influence the way language is structured and most important its underlying meaning in the spatial context. This is in good agreement with Vygotsky's (1993) investigation of the relationship between language and thinking, whose union is the meaning of words representing "such a close amalgram of thought and language that it is hard to tell whether it is a phenomenon of speech or a phenomenon of thought" (Vygotsky, 1993, p. 212) (see Section 2.2.2). Although a deeper understanding of the word meaning in spatial discourse requires an analysis of spatial language, this does not necessarily imply that the underlying phenomena are situated only within the "artificial" boundary of language as a communicative medium, because the reconstruction of word meanings requires spatial knowledge and spatial abilities. Such an observation complies with assumptions from previous works (e.g., Maier & Schweiger, 1999; Wessel, 2015) about the interconnected communicative and cognitive functions of language in mathematics classroom (see Section 2.2.3).

Moreover, language can be considered as structuring the reconstructed spatial knowledge, which can be observed in interpretation of spatial language from the builder's perspective in the reconstruction method. Another important issue regarding the influence of language on spatial abilities as manipulation and transformation of objects in space is the analysis of reconstructed spatial objects, which spatial construction is primarily mediated by language. By analysing the builder's manipulating actions on spatial objects, the researcher can reconstruct the builder's interpretation of communicated spatial knowledge and observe the influence which language exercises on the space in context. Although this phenomenon arises from the exceptional situation in the reconstruction method, such unique cases in research methodology may be fruitful for investigating and shedding light on the interplay between language and spatial ability.

6. Results and discussion from the deductive data analyses

This chapter is dedicated to the explorative research perspectives arising from the application of quantitative methods to the student spatial language collected using qualitative methods in this present research. The aim of this chapter is to investigate the frequency of use of identified strategies in student discourse and the nature of spatial language under consideration of possibly influencing background factors – student language proficiency, spatial abilities, and sex. The results to corresponding research questions (R3) and (R4) in Section 3.1 were achieved by the deductive data analysis processes described in Section 4.2.6.3. In Section 6.1, the use of strategies identified in Section 5.1 is investigated under consideration of the above-mentioned background factors. The next section, Section 6.2, is dedicated to the structural analysis of students' spatial language emerging in the solving of the underlying spatial-verbal tasks, whereby the structure of student spatial language is analysed under consideration of the students' language proficiency, spatial abilities, and sex.

6.1 Use of identified strategies under consideration of influencing factors

After having investigated different strategies used and obstacles encountered by students whilst describing the spatial objects on a qualitative level, the use of the strategies identified in Section 5.1 is now analysed in consideration of students' background factors. As previously described in the theoretical sampling (see Section 4.2.3.1), three factors were considered as possibly influencing the use of the identified strategies: students' language proficiency, spatial abilities, and sex. The use of the identified strategies is analysed in consideration of the three dichotomies induced by these three factors, which is formulated by the third research question:

(R3)   Does the use of the identified strategies vary with the language proficiency, spatial abilities, and sex of the students?

It is plausible that the number of participants in the study is a limitation regarding the investigation of significant relationships between use of identified strategies and the above-mentioned background factors, hence the following findings should be treated with caution. Nevertheless, it is interesting to explore the distribution of use of identified strategies among the student groups based on the

175

theoretical sampling. The diversity of the sampling group (see Figure 14) and the underlying observations and patterns regarding the strategy use should shed light on possible relationships based on hypotheses from previous literature about solving spatial and verbal tasks. Such an explorative approach to the use of identified strategies can act as a basis for future quantitative research about the possible influence of such factors on solving spatial tasks using spatial language. In the following sub-sections, the hypotheses about predicted use of identified strategies based on previous findings and the actual findings based on statistical analyses and sorted according to the influencing factors are described.

6.1.1 Hypotheses

In this sub-section, hypotheses about the expected outcome of the observation of the use of identified strategies considering students' language proficiency, spatial abilities, and sex are formulated. A hypothesis is a statement which explores a prediction about the relationship between two or more variables and is suggested by underlying observations and knowledge, which requires testing for validation or falsification (cf. Creswell, 2003). The hypotheses are based on and justified by results from previous studies about spatial abilities, language, and sex differences. The hypothesis about the use of strategies will be sorted according to the underlying influencing factor, and tested using simple statistical analyses in the upcoming sections.

Hypothesis 1 (H1): Students with high language proficiency are expected to use more identified strategies in their spatial discourse than students with low language proficiency.

The factor, language proficiency, should play a major influence on the use of identified strategies, since identified strategies must be realised by the medium of language. Hence, students with high language proficiency should be in a better position to verbalise their spatial thinking, assuming that the C-test scores reflect the student's high language fluency and accuracy (especially in the ability of speaking and communicating knowledge – including language awareness and adaptiveness) and vocabulary acquisition. Hence, students with high proficiency should be able to verbalise more their spatial thinking in their discourse, hence they are expected to use more strategies in their discourse. In particular, analytic strategies should be used more by students with high language proficiency, because the recognition of charactersitics and properties tend to be expressed using concepts (originating in the verbal information processing cognitive system), which are more viable for students with a more developed language register (cf. Barrat, 1953; Schwank, 2003 in Section 2.3.2). A case which shows the importance of language proficiency for the use of strategies is represented at the beginning of Section 5.2.3, whereby Describer 9 with low language proficiency and high spatial abilities was not able to verbalise the rotating actions, hence the rotating strategy could not be identified in his spatial discourse. Hence, students with low language proficiency are expected to use less identified strategies due to their less developed language competencies.

Hypothesis 2 (H2): Students with high spatial abilities are expected to use more identified strategies, especially holistic strategies, in their spatial discourse than students with low spatial abilities.

Given the spatial nature of the task, which is represented in the purpose for which language is used in the task solving process, spatial abilities should also be an influencing factor for the use of identified strategies. Although all identified strategies represent the structuring and communication of spatial knowledge, the verbal-holistic strategies of break-down, assembling and rotation seem to require more spatial manipulation and mental actions on spatial objects rather than the analytic strategies (see Sections 2.1.5.2 and 5.1.7). The consideration of holistic strategies as being more "pure" mental-visual spatial ability strategies in previous literature (e.g., Barratt, 1953, Schulz, 1991) supports the expectation that students with higher spatial ability (which is measured by instruments based on a psychological model of spatial ability; see Sections 2.1.3.2 and 3.5.3.2.2) tend to make use more of these strategies (in contrast to analytic strategies). Hence, students with high spatial abilities are expected to use more from all the identified strategies, especially from the verbal-holistic strategies mentioned above.

Hypothesis 3 (H3): Female students are expected to use more identified strategies, especially analytic strategies, in their spatial discourse than male students.

In Section 2.1.6, results from different studies about sex differences and solving of spatial and verbal tasks were discussed. Various studies (e.g., Rost, 1977; Linn & Petersen, 1985; Burton et al., 2010; Geiser et al., 2006) indicate that males show better performance in mental rotation and spatial orientation, whereas females show better performance in verbalising skills when compared to males. Based on results of such studies, which show the predominant use of holistic and analytic strategies among males and females respectively (e.g., Coluccia et al., 2007), male students are expected to use more holistic strategies and females more analytic strategies, considering the assumption that females are better at verbalising.

After having formulated and justified the three hypotheses based on the influencing factors, the results concerning the use of strategies among the describers in the reconstruction method are presented followed by the testing of the hypotheses in the upcoming sections accordingly.

6.1.2 Language proficiency as a possible influencing factor

For the testing of hypotheses formulated in Section 6.1.1, an overview of the absolute frequency of use of strategies in the sixteen describers' spatial discourse is illustrated in the following table:

Student	M	BD	A	R	CC	SC	Number of identified strategies
D-1	6	3	1	0	0	0	10
D-2	5	3	2	0	0	0	10
D-3	7	3	2	1	0	4	17
D-4	7	2	2	0	0	0	11
D-5	5	3	0	0	0	0	8
D-6	10	3	2	0	0	0	14
D-7	6	4	2	0	0	1	13
D-8	12	3	2	1	0	2	20
D-9	6	2	1	0	0	2	11
D-10	10	3	2	1	3	0	19
D-11	6	2	1	0	0	0	9
D-12	8	3	2	0	0	0	13
D-13	3	3	1	0	0	0	7
D-14	2	2	0	0	0	0	4
D-15	4	3	2	0	0	0	9
D-16	8	2	0	0	0	0	10
Median							10,5

Table 21   Results about the use of different strategies by the sixteen describers in their spatial discourse in the reconstruction method

For testing of hypothesis H1, the results about the absolute frequency of use of the identified strategies are represented dependent on two groups: eight students with high language proficiency and eight students with low language proficiency. The results of the use of *spatial metaphors* (M), *breakdown* (BD), *assem-*

bling (A), *rotation* (R), *counting controlling* (CC), and *structure controlling* (SC) strategies in the describers' spatial discourse when solving the spatial task in the reconstruction method are visualised in the following table:[56]

Language Proficiency	n[57]	M	BD	A	R	CC	SC
High	8	58 (55%)	24 (55%)	13 (59%)	2	0	7
Low	8	47 (45%)	20 (45%)	9 (41%)	1	3	2

Table 22 Absolute frequency of use of identified strategies regarding students' language proficiency[58] [59]

The results illustrated in Table 22 show the absolute frequency of different realisations (or elements) of strategies employed in the students' spatial discourse, not considering their repetitive use in the same spatial discourse. It is important to note that the repetitive use of same strategies refers to different elements of strategies, rather than the strategy as a group. There are different possibilities of using spatial metaphors. All these different realisations of spatial metaphors were coded separately. For example, the spatial metaphors of *ladder* and *H* have been coded as two different spatial metaphors, but how much these specific metaphors were used in discourse was not coded in this section. The results in Table 22 reveal that spatial metaphors are the most common strategy used by the students to solve the spatial task in the reconstruction method. It is important to note that the values for spatial metaphors in Table 22 represent different realisations of the spatial metaphors among different students. Hence, the first value denotes 58 spatial metaphors, which are unique in student's spatial discourse of the underlying group, but this value includes the multiple use of the same metaphors among different students. Different students might have used the same spatial metaphors in their spatial discourse, as, for instance, the spatial metaphor *staircase*, which was used by fourteen students to describe the structure of spatial object A. Nevertheless, the high use of spatial metaphors in general might imply the high analytic nature of the spatial task, which requires the description of various spatial properties. Moreover, Table 22 shows that the spatial meta-

---

[56] As described earlier in the data analysis section, spatial tasks A and B should be considered as one spatial task for data analysis purposes.
[57] Number of students (describers).
[58] A numeric representation of the results was considered as more adequate, because of the relatively high occurrences of the first strategy in contrast to the other strategies.
[59] This representation of the results focuses on the distribution between the two groups, rather than on use among the different students within both groups.

phors were slightly used more by students with high language proficiency, i.e. 10 percentage points (p.p.) more, in contrast to students with low language proficiency. A similar finding can be observed in the use of breakdown and assembling strategies, which was used 10 p.p. and 18 p.p. more by students with high language proficiency respectively. In contrast to the first three strategies, rotation strategy was only used three times, by two students with high language proficiency and by one student with low language proficiency. Interestingly, the use of cubes controlling strategy was observed only in the group of students with low language proficiency, more precisely, it was used three times only by one student with low language proficiency in his spatial discourse. In the second category of controlling strategies, structure controlling strategy was used multiple times, i.e. seven times, among three students with high language proficiency and twice by a student with low language proficiency. Due to the fact that these findings are only preliminary observations, this data was analysed using statistical descriptive analysis methods, i.e. contingency tables and chi-squared tests of independence, for an explorative approach for testing the hypotheses formulated in Section 6.1.1. The following contingency tables illustrates the frequency distributions of the use of strategies (high or low using median split) and the students' language proficiency. Whereas Table 23 shows that frequency distribution of use of all strategies and language proficiency, Table 24 and Table 25 illustrate the frequency distribution of student language proficiency and the use of analytic and holistic strategies respectively.

	Language proficiency		Total
Use of strategies	High	Low	
High	5	3	8
Low	3	5	8
Total	8	8	16

Table 23 Contingency table of frequency distribution of the variables students' use of strategies and language proficiency

	Language proficiency		Total
Use of analytic strategies	High	Low	
High	5	4	9
Low	3	4	7
Total	8	8	16

Table 24 Contingency table of frequency distribution of the variables use of analytic strategies and language proficiency

Use of holistic strategies	Language proficiency		Total
	High	Low	
High	7	4	11
Low	1	4	5
Total	8	8	16

Table 25   Contingency table of frequency distribution of the variables use of holistic strategies and language proficiency

	Value	df	Asymptotic significance (2-tailed)	Exact sig-nificance (2-tailed)	Exact signifi-cance (1-tailed)
Chi-squared from Pearson	2.618	1	0.106		
Correction for continuity	1.164	1	0.281		
Likelihood-Quotient	2.756	1	0.97		
Exact Test from Fisher				0.282	0.141
Number of valid cases	16				

Table 26   Chi-squared test of independence on a 2 x 2 contingency table of frequency distribution of the variables use of holistic strategies and language proficiency

Following a computation of chi-sqaured tests of independence for the above contingency tables, no significant relationship could be observed between the variables student language proficiency and use of strategies in general ($P = 0.317 > 0.05$), language proficiency and use of analytic strategies ($P = 0.614 > 0.05$), and language proficiency and use of holistic strategies ($P = 0.106 > 0.05$) (see Table 26). Even though first observations of data in Table 22 have shown a slightly higher use of identified strategies among students with high language proficiency, the use of identified strategies does not differ statistically significantly among students with high language proficiency and students with low language proficiency. Hence, the findings do not support the hypothesis formulated in H1.

6.1.3 Spatial abilities as a possibile influencing factor

As indicated in hypothesis H2, students' spatial abilities should be another promising factor which can play an important role in developing the identified strategies in spatial discourse. Similar to Table 22, Table 27 visualises the distribution of use of identified strategies according to two dichotomies: eight students with high spatial ability and eight students with low spatial ability.

Spatial abilities	n	M	BD	AS	R	CC	SC
High	8	55 (52%)	21 (48%)	13 (59%)	2	3	6
Low	8	50 (48%)	23 (52%)	9 (41%)	1	0	3

Table 27   Absolute frequency of use of identified strategies regarding students' spatial abilities

In contrast to the distribution of identified strategies according to students' language proficiency in Table 22, the gaps between the use of spatial metaphors and break down strategies among students with high and low spatial abilities are less remarkable (see Table 27). In contrast, a higher use of the assembling strategy (and to a more limited extent also rotation and cube controlling strategy) can be noticed among students with high spatial abilities. Similar to the results under consideration of language proficiency, the stategy of rotation was used by two students with high spatial abilities and one student with low spatial abilities in their spatial discourse. It might be important to highlight that these two students with high spatial abilities scored very high scores in the mental rotation test in the pencil-and-paper tests, which were used for spatial ability assessment in sampling (see Section 4.2.3.2.2). However, due to the small size of the sample group and limited use of this strategy it is not possible to make any statements about the relationship between the strategy rotation and scores in mental rotation tests. However, one could put up the hypothesis that there might be a link between both and investigate it in future research.

With regard to the expected higher use of verbal-holistic strategies in the spatial task solving process, the only remarkable higher use can be noticed among strategies of assembling and rotation, whereas the break down strategy is interestingly used slightly more by students with low spatial abilities. It is important to note that this does not reflect the successful use, since some students might have applied different break-down strategies on the same object, especially due to the existence of the control moment in spatial task A.

Use of identified strategies	Spatial abilities		Total
	High	Low	
High	5	3	8
Low	3	5	8
Total	8	8	16

Table 28 Contingency table of the frequency distribution of the variables spatial abilities and use of identified strategies

In order to test the hypothesis H2, contingency tables and their corresponding chi-squared tests were developed to investigate the relationship between students' spatial abilities and their use of strategies. The contingency table in Table 28 shows that the proportion of students with high spatial ability using a high number of identified strategies is the same as the proportion of students with low spatial ability using a low number of identified strategies. However, Pearson's chi-squared tests of these values show that the use of identified strategies does not vary significantly (P = 0.317). Similarly, no significant variation was observed among the use of holistic (P = 0.590) or analytic (P = 0.614) strategies and students' spatial ability. Hence, results from the output from the statistical hypothesis tests do not support the hypothesis formulated in H2, i.e. students with higher spatial abilities do not significantly use more identified strategies than students with low spatial abilities in their spatial task solving process.

6.1.4 Sex as a possible influencing factor

The third factor, sex, can also be an influencing factor for solving spatial tasks, as earlier research on spatial abilities has proven (see Section 2.1.6). Therefore, the use of identified strategies should be analysed under consideration of two groups: eight female students and eight male students. Table 29 shows the results for the frequency of use of strategies based on the sex dichotomy:

Sex	n	M	BD	A	R	CC	SC
Male	8	47 (45%)	22 (50%)	9 (41%)	1	3	7
Female	8	58 (55%)	22 (50%)	13 (59%)	2	0	2

Table 29 Absolute frequency of use of identified strategies regarding students' sex

The results in Table 29 show that females tend to use more spatial metaphors, assembling, rotation and structure controlling strategies in their spatial discourse, which shows similar patterns to the results for students with high language proficiency in Table 22. However, no difference is noticed between the use of breakdown strategy among females and males. With reference to analytic and holistic strategies, the above results can be interepreted as showing a tendency of higher use of verbal-analytic strategies among female students. This data was used for further investigation of any possible significant difference regarding use of analytic strategies between male and female students.

	Value	df	Asymptotic significance (2-tailed)	Exact significance (2-tailed)	Exact significance (1-tailed)
Chi-squared according to Pearson	2.286	1	0.131		
Correction for continuity	1.016	1	0.313		
Likelihood-Quotient	2.348	1	0.125		
Exact Test from Fisher				0.315	0.157
Number of valid cases	16				

Table 30 Chi-squared test of independence on a 2 x 2 contingency table of frequency distribution of the variables use of analytic strategies and sex

Upon computation of the chi-squared hypothesis tests for the contingency tables of the two variables, no significant relationship could be observed between the variables of sex and the use of analytic (P = 0.131) (see Table 30) or holistic (P = 0.590) strategies in describers' spatial discourse. The first P-value shows that a certain trend toward significance might be interpreted to a certain extent, however, under assumption of a significance level of 0.05, the hypothesis that female students use more of identified strategies, especially analytic strategies, cannot be statistically significantly validated.

6.1.5 Discussion of the use of identified strategies

Before discussing the results observed in the above three sub-sections, the results should be visualised according to the four different sample groups, A (high language proficiency and high spatial abilities), B (high language proficiency and low spatial ability), C (low language proficiency and high spatial abilities), and D (low language proficiency and low spatial abilities), which were established for the sampling (see Section 4.2.3.1), in order to better understand the distribution of the use of identified strategies among the participating students (see Table 31).

Group	n	M	BD	AS	R	CC	SC	Total
A	4	25	11	7	1	0	4	49 (26 %)
B	4	33	13	6	1	0	3	56 (30 %)
C	4	30	10	6	1	3	2	52 (28 %)
D	4	17	10	3	0	0	0	30 (16 %)

Table 31   Frequency of use of identified strategies sorted according to the four sample groups

Table 31 shows the distribution of use of the identified strategies concerning the factors of students' language proficiency and spatial abilities. The results show that students with low spatial abilities and low language proficiency (Group D) make less use of the identified strategies, when compared to the other students from other groups. In particular, the use of spatial metaphors, assembling strategy, rotation and controlling strategies was more common among students with high language proficiency or high spatial abilities or both. The higher use of controlling strategies among students with either high language proficiency or high spatial abilities or both are in line with results from previous studies (e.g., Zimmerman & Pons, 1986), which show that high achieving students display significantly greater use of self-regulating strategies. For a deeper investigation of this

184

observation, I would like to illustrate the use of identified strategies in a case study of a student with low language proficiency and with low spatial abilities in the following transcript:

Transcript 24:

(Describer 14 got a glimpse of the rebuilt object during the control moment (see below), is instructed to redescribe spatial object A to the builder).

	(...)	*(...)*
Describer 14:	*Firstly, four stones in a row.*	*Erstmal vier Steine in einer Reihe.*
Builder 14:	*Yes.*	*Ja.*
Describer 14:	*On it, three in the right one... and the left one must stay empty, the left stone.*	*Darauf drei in die rechte... und das linke muss frei bleiben, der linke Stein.*
Builder 14:	*Yes.*	*Ja.*
Describer 14:	*Two on it, and the left one stays free again.*	*Darauf zwei, und der linke bleibt wieder frei.*
Builder 14:	*Yes.*	*Ja.*
Describer 14:	*And then one stone, and the left one stays empty again.*	*Und dann ein Stein, und das linke bleibt wieder frei.*
Builder 14:	*Yes.*	*Ja.*
Describer 14:	*Now you build one at the topmost stone, to it at the side.*	*Jetzt baust du an dem obersten Stein eins dran an*
Builder 14:	*Left or right?*	*Links oder rechts?*
Describer 14:	*Left. (...) Under it, under the left stone there are two again. (...) Under it, there are three under that you have just built. (...). And unter it four and the three which you have just built. (...) That's it.*	*Links. (...) Darunter unter dem linken Stein kommen wieder zwei. (...). Darunter kommen drei unter die du grad gebaut hast. (...). Und darunter vier und die drei die du grad gebaut hast.*

The above excerpt shows a student from Group D after the control moment in spatial task A. One can notice the few strategies which are used by Describer 14 in his description of the spatial object, especially the lack of use of spatial metaphors, breakdown of spatial object in two internal parts, assembling, rotation or control strategies. Such a case indicates the tendency to support the hypothesis that low language proficiency and low spatial abilities might be the reason for the relatively low use of strategies in the solving of the spatial task. The spatial information of the spatial object requires Describer 13 to develop a strategy for emphasising and communicating the three-dimensional front view of spatial object A, which requires a strong simultaneous command of language and spatial abilities in the spatial description. In this case, Describer 14 chose to rely consciously on conventional spatial elements, i.e. spatial prepositions, considered to be a task demand, and developed relatively fewer strategies in his discourse. However, from a qualitative approach, it is not clear whether the lower scores in language proficiency and in spatial abilities are explicitly related to this phenomenon, i.e. the lower use of strategies in spatial discourse.

Despite of the small sampling size, preliminary results indicate an inclination to the three formulated hypotheses from Section 5.3.1 to a certain extent: students with high language proficiency or with high spatial abilities or females tend to use more identified strategies than others. However, the preliminary results concerning the males' and females' preference of verbal-holistic and verbal-analytic strategies should be interpreted with care, since, for instance, females have used more strategies – whether of holistic or analytic nature – than males. Therefore, this finding does not confirm the preference of holistic or analytic strategies among females or males accordingly, for instance, as indicated by Coluccia et al. (2007). Apart from this slight non-alignment, the preliminary results could show an inclination to the indication of females' (independently from their language proficiency) better verbal fluency, which is an important prequisite for every strategy use in the reconstruction method, compared to the male participants. However, all results of the statistical hypothesis tests showed no significant relationship between the use of the identified strategies, whether analytic or holistic in nature, and the each of the three variables: student language proficiency, spatial abilities, and sex.

Another remarkable observation to emerge from the data comparison is the fact that no substantial difference about use of identified strategies could be noticed among students in Groups A, B and C (see Table 31). Nevertheless, the finding about the contrast to Group D can imply the preassumption that both language proficiency and spatial abilities seem to be promising influencing factors which play an important role when solving spatial-verbal tasks. However, given that the findings are based on a limited number of participants and tasks, the results from such analyses should consequently be treated with considerable caution. Again, the use of the identified strategies was analysed independently from their success in the task solving process, hence a use of an identified strategy should not necessarily be viewed as advantageous, as it wasdiscussed in possible obstacles, for instance, of spatial metaphors in Section 5.2.1. Nevertheless, the use of different strategies and their flexible or adaptive use play an important role in the development of strategic ability in solving problems, especially in spatial tasks (cf. Lemaire & Siegler, 1995). Hence the different use of strategies and their flexible use by students with high spatial abilities or high language proficiency or both should reassert the more developed strategic abilities which could to be related to the higher performances in the instrument tests. Moreoever, the findings about similar patterns in use of identified strategies between students with high language proficiency and/or high spatial abilities can be interpreted as a confirmation of the interweaving of spatial and verbal competencies in underlying spatial task. The finding that Group D made a substantially less use of identified strategies (see Table 31), can presumably imply that the development of strategic competence in the spatial solving task requires either a strong grasp of language or of spatial knowledge or both. This observation can be used to indicate the strong connection between language and spatial abilities (as content learning) in mathematics education.

Whereas, the use of identified strategies shows how high and low achieving students approach spatial-verbal tasks, it does not reveal much about the nature of spatial language which is used by the describers to achieve the goal of the task. In the next section, Section 6.2, I will provide various structural approaches to analyse spatial language of describing students with different achievements and sex.

6.2 Deductive approaches for structural analysis of spatial language and influencing factors

In this section, I will provide structural approaches to the analysis of spatial language. Based on the assumption that spatial language provides an insight to students' spatial thinking (cf. Levinson, 1996; Landau & Jackendoff, 1993), if students' spatial abilities varies, then so should their spatial language. Therefore, it is interesting to analyse the spatial language of students under the consideration of influencing factors of language proficiency, spatial abilities, and sex. Hence, this section provides findings to the following research question:

(R4) To which extent can spatial language be analysed structurally and does its use differ among students with different background factors?

Up to this section, spatial language was described as the language used to describe spatial objects, their spatial position in space, the spatial relations between different spatial objects, and the language used for solving verbal-spatial tasks. In particular, findings about the strategies used and obstacles encountered by students when solving spatial tasks, in Section 5.1 and Section 5.2 respectively, support the need to extend the notion of spatial language by the language used for externalising spatial thinking. Hence, a notion of spatial language should also include properties such as describing transformations of spatial objects, such as movement of objects in space, and not just their static spatial properties. Such a notion of spatial language is realised in solving of spatial tasks in the reconstruction method, where both holistic and analytic strategies need to be verbalised in the solving processes (see Section 5.1.7).

After determining the strategies and obstacles which may influence students' spatial discourse in the reconstruction method, a more structural approach of the student spatial language is needed to understand its development among students with different achievements and sex. Therefore, the last research question aims to consider features of students' actual spatial language and their quantification under consideration of the different influencing factors mentioned above. This is done by providing four different approaches for analysing student spatial language structurally: metaphoric-based approach, linguistic-based approach, content-based approach and conception-based approach. These approaches are described in detail in the following sections.

6.2.1 Metaphoric-based approach

As mentioned earlier in this present research, spatial metaphors are important elements of spatial language and can be analysed by considering different dimensions: the linguistic, the spatial and the conceptual dimension (see Section 5.1.1). Therefore, an analysis of spatial metaphors in discourse should provide insights into students' metaphorical thinking in space. In particular, I would like to investigate whether the type of spatial metaphors used by the describing students in their spatial discourse differ when considering their language proficiency, spatial abilities, and sex. In contrast to Section 6.1, in which the absolute frequency of use of different spatial metaphors in students' spatial discourse was in focus, the multiple occurrences of the spatial metaphors (in the describer student discourse) in relation to the length of student spatial discourse (which was analysed in terms of total number of phrases, a phrase being a group of words which form a conceptual unit and a clause), i.e. the relative frequency, was considered.[60] Before describing the results, the following hypotheses are considered for the expected use of spatial metaphors in the students' spatial discourse in the reconstruction method.

Hypothesis 4 (H4): Female students, or students with high language proficiency, or students with high spatial abilities, are expected to use more spatial metaphors in their spatial discourse than other students.

Referring to the findings presented in Section 6.1, spatial metaphors were slightly used more by female students, students with high language proficiency or with high spatial abilities. Hence, we expect their spatial language to be more structured by the use of spatial metaphors, in contrast to other students.

Hypothesis 5 (H5): Students with high language proficiency are expected to use more mathematical spatial metaphors in their spatial discourse than students with low language proficiency.

Assuming that students with high language proficiency have a wider vocabulary knowledge, this group of students are expected to use more mathematical spatial metaphors in their spatial language. Furthermore, this hypothesis is supported by fact that language is an important factor for mathematics learning, hence for learning and higher use of mathematical concepts (cf. Prediger et al., 2013).

Hypothesis 6 (H6): Students with high spatial abilities are expected to use more spatial metaphors with diverse functions than students with low spatial abilities.

Apart from analysing the linguistic aspects of spatial metaphors, the underlying meaning of metaphors used in spatial discourse should also be investigated. In Section 5.1.1, three functions of metaphors were identified and described, which reflected some of the cognitive processes of visual perception demanded in spatial tasks (see Section 2.1.2.1). Students with high spatial abilities are expected to

---

[60] For the relative frequency, the absolute use of spatial metaphors in discourse is set in relation to total number of phrases, in which the student's discourse have been broken down into.

show a more diverse use of metaphors with different functions – structure, spatial position and spatial relation functions – compared to students with low spatial abilities. Before testing the above-mentioned hypothesis in the following sections, I would like to present the results of the distribution of use of spatial metaphors in the spatial discourses of the sixteen describers in Figure 42.

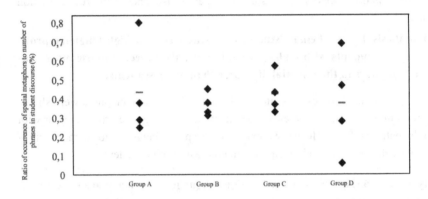

Figure 42 Distribution of use of spatial metaphors in spatial discourse among the sixteen students sorted according to sample groups

On average (represented by the horizontal grey lines in Figure 42), no substantial difference of use of spatial metaphors in spatial discourse according to the sample groups can be observed. One can notice two observations: an outlier[61] in group A and relatively wide dispersion of the values in group D. Under consideration of these observations, the findings of use of spatial metaphors according to the linguistic, spatial and conceptual dimension are introduced and discussed in the upcoming sections.

6.2.1.1 The linguistic dimension of spatial metaphors and influencing factors

In the linguistic dimension, spatial metaphors were categorised in three groups according to the type of language used: everyday language, letter-based language, and mathematical language (see Section 5.1.1.1). Table 32 shows the average relative frequency of occurences of metaphors in students' spatial discourse, which is a percentage made up of the ratio of actual occurrences of these

---

[61]   This high value has been achieved due to a high number of spatial metaphors in a relatively short spatial discourse in the reconstruction method.

components to the total number of phrases in the describing student's spatial discourse sorted according to the four sample groups (see Table 32).[62] For instance, the value 36% in the first cell in Table 32 indicates that on average students in Group A used everyday spatial metaphors in 36 percent of their average number of phrases in their spatial discourse.

Linguistic dimension	Group A	Group B	Group C	Group D
Everyday	36%	25%	36%	16%
Letter-based	7%	5%	3%	1%
Mathematical	4%	7%	2%	6%

Table 32   Average relative frequencies of occurrences of spatial metaphors in student spatial discourse among the sample groups under the consideration of their linguistic dimension[63]

Table 32 shows that spatial metaphors from everyday language tend to be used on average more by students who scored high in language proficiency and spatial tests (approximately one in every third phrase), whereas participants from Group D have used the least number of spatial metaphors in their spatial description (approximately less than one in every fifth phrase). The following table represents the average ratio of occurrences of the different spatial metaphors (according to the linguistic dimension) to the average number of phrases of students with high or low language proficiency (see Table 33).

Linguistic Dimension	High language proficiency	Low language proficiency
Everyday	30%	26%
Letter-based	6%	2%
Mathematical	6%	4%

Table 33   Average relative frequencies of occurrences of everyday, letter-based and mathematical spatial metaphors in student spatial discourse under consideration of their linguistic proficiency

From the above data, one can only notice minor differences between the type of language used among students with high and low language proficiency. The relatively small gap of 2 p.p. is not promising enough to induce a statistical hypothesis test to provide answers for hypothesis 5, i.e. that students with high language proficiency use relatively more mathematical spatial metaphors. However, in

---

[62]   The spatial discourse of describers was dismantled in utterances for this purpose.
[63]   The percentage is based on the ration of number of intended spatial metaphors in relation to the total number of phrases in the students' spatial discourse.

general, both students with high language proficiency or high spatial abilities or both tended to use more different spatial metaphors (especially everyday spatial metaphors) in their spatial discourse than students with low language proficiency or low spatial abilities or both (see Table 32). If one considers the use of language in spatial metaphors among male and female students, one can notice that females used slightly more spatial metaphors from the mathematical language and less from everyday language in their spatial discourse when compared to males (see Table 34).

Linguistic dimension	Male	Female
Everyday	31%	25%
Letter-based	6%	8%
Mathematical	3%	6%

Table 34   Average frequencies of occurrence of everyday, letter-based, and mathematical spatial metaphors in students' discourse under consideration of their sex

This preliminary finding appears to concur with the results of other studies (e.g., Rost, 1977; Geiser et al., 2006), which indicate that female students show a more developed language acquisition when compared to their male counterparts. However, careful attention should be paid in this interpretation, because compared to the values for everyday language spatial, the values for mathematical spatial metaphors are relatively low. Further analyses of data were performed using descriptive statistics tests to test for any significance, such as the nonparametric Mann-Whitney U test. The Mann-Whitney U test was considered as adequate for such purpose, because it is a nonparametric test used for experiments in which there are two conditions and different subjects were used in each condition. Results from the Mann-Whitney U tests showed no significant difference in the use of the different types of spatial metaphors in the linguistic dimension and students' background factors. For instance, the result $P = 0.958$ was achieved upon calculation of Mann-Whitney-test for the variables everyday spatial metaphors and language proficiency.

6.2.1.2 The spatial dimension of spatial metaphors and influencing factors

In the spatial dimension, spatial metaphors were categorised according to their function in spatial terms. Spatial metaphors can be used to describe the structure of spatial objects, their spatial position or the spatial relation between two spatial objects (see Section 5.1.1). Table 35 represents the average relative frequency of use of the spatial metaphors according to the three functions in relation to the average number of phrases in spatial discourse of the four sample groups.

Spatial dimension	Group A	Group B	Group C	Group D
Structure (ST)	38%	31%	34%	36%
Spatial position (SP)	3%	4%	8%	0%
Spatial relations (SR)	2%	2%	0%	2%

Table 35 Average frequencies of occurrence of spatial metaphors according to their functions among the four sample groups

As expected, most spatial metaphors were used by students to describe the structure of the spatial objects, with relatively small portions being used to describe the spatial position or spatial relations in their spatial discourse. In relation to spatial relations, overall, spatial metaphors were relatively used more to describe the spatial position. However, one can notice that Group C and Group D did not make use of spatial metaphors for spatial relations and spatial position respectively. Especially, spatial metaphors for spatial position is more likely not to be used among students in Group D given the higher differences to the scores of the other three groups in Table 35. The results in Table 35 illustrate the relatively rare use of spatial metaphors with SR-function, which might be accounted for by the fact that other linguistic means can be used for this purpose, for example, spatial prepositions. Consider the functions of the used spatial metaphors under consideration of the the students' spatial abilities. Table 36 represents which functions of spatial metaphors were used most among students with high or low spatial abilities.

Spatial dimension	High spatial abilities	Low spatial abilities
Structure (ST)	36%	34%
Spatial position (SP)	6%	2%
Spatial relations (SR)	1%	2%

Table 36 Average frequencies of occurrence of spatial metaphors according to their function in students' spatial discourse according to their spatial abilities

The noteworthy difference between the functions of spatial metaphors used between the groups is that students with high spatial abilities show a higher tendency of using spatial metaphors to describe spatial positions than students with low spatial abilities, even though the values are relatively low when compared to the values of ST-function. Regarding the other factors, language proficiency and sex, no substantial differences could be observed regarding the functions of spatial metaphors used in the students' spatial discourses. Therefore, even though the values are not very high, the results can be interpreted as showing a tendency that students with high spatial abilities make more use of spatial metaphors with structure and spatial position in their spatial discourse. Hence, statistical hypothesis tests were applied to the data to determine whether there is any statistically significant difference between use of spatial metaphors with spatial position function and spatial abilities. Table 37 shows an example of the results from a Mann-Whitney U test, which shows that there is no significant difference between the use of the SP-function among students with low and students with high spatial abilities.

	SP
Mann-Whitney U	21.000
Wilcoxon W	57.000
Z	-1.178
Asymptotic Significance (2-tailed)	0.239
Exact significance [2*(1-tailed Sig.]	0.279

Table 37   Mann-Whitney U test for testing the difference of use of spatial metaphors with SP-function among students with high and low spatial ability

6.2.1.3 The conceptual dimension of spatial metaphors and influencing factors

In the conceptual dimension of spatial metaphors, spatial metaphors are categorised according to their nature of mathematical thinking in conceptual terms according to Sfard (1999). Spatial metaphors can be used to describe a static spatial configuration or a configuration which requires dynamic thinking or movement, as it was explained in Section 5.1.1. Table 38 presents the results of the type of conceptions of spatial metaphors used by the students when solving the verbal-spatial task in the reconstruction method.

Conceptual Dimension	Group A	Group B	Group C	Group D
Static (S)	42%	34%	39%	38%
Dynamic (D)	1%	3%	3%	3%

Table 38   Use of static and dynamic spatial metaphors among the four sample groups

The results in Table 38 show that the majority of the spatial metaphors used by the students are of static nature, and just a few are dynamic. In general, students seem to show preference of using metaphors of static nature, which seems to be linked to the finding that students (independent from background factors) use more spatial metaphors with ST-function (see Section 6.2.1.2). The findings in the last sections show to what extent spatial metaphors are used to structure spatial language for solving the spatial task(s) in the reconstruction method. Whereas the factors of language proficiency did not prove itself as substantially influencing, the factors of sex and spatial abilities have proven to be more remarkable when analysing the linguistic and the spatial dimension of spatial metaphors in discourse respectively. However, all descriptive statistics showed no significant difference between use of the different types of spatial metaphors, whether in their linguistic, spatial or conceptual dimension, and the student background factors. Nevertheless, the above results indicate which types of metaphors students most likely use in their spatial language, in order to solve spatial tasks in the reconstruction method, whereby the most common spatial metaphors are realised using everyday language for describing the structure statically. However, spatial metaphors were not the only elements carrying spatial meaning and structuring spatial language, as the following sub-section about a linguistic-based approach (excluding spatial metaphors) for analysing spatial language, Section 6.2.2, demonstrates.

6.2.2 Linguistic-based approach

Beside spatial metaphors, there are more characteristic features of spatial language, which can serve to analyse student's spatial language structurally for achieving a deeper understanding of the type of language used in spatial discourse. In particular, elements carrying spatial meaning in spatial discourse, excluding spatial metaphors (already analysed in the metaphoric approach to spatial language in Section 6.2.1), is considered for this purpose. From a linguistical point of view, such elements can be categorised in two groups: spatial prepositions and spatial adverbs (without spatial deixis).[64] Spatial prepositions can be used to describe the spatial relation between two or more objects, which is reflected in the need of a figure and ground in discourse. In contrast, spatial adverbs focus more on the location of the object, so where the action of a verb is carried out. The following table, Table 39, presents a linguistic-based categorisation of elements carrying spatial meaning in spatial discourse (in the German language, as the data collected in this study), by giving an overview of the spatial prepositions and adverbs used by describing students in their spatial discourse when solving the spatial task. It is important to note that whereas the German prepositions and adverbs are distinctive, the classification of language elements in English spatial prepositions and averbs is not always unique (in English the

---

[64] Due to the spatial disambiguity of use of spatial deixis in the reconstruction method and due to the need to predefine them in context, spatial deixis should not be included in the set of spatial adverbs for this linguistic analysis of spatial language.

distinction between adverb and preposition requires an analysis of the word following the element and hence cannot be easily classified as in the German language).

Spatial prepositions	Spatial adverbs
Außerhalb, zwischen, vor, unter, über, mittig von, in, hinter, auf, gegenüber, an/bei, neben.	*Oben, unten, rechts, links, hinten, vorne.*
Outside of, between, in front of, under, over, middle of, in, behind (of), on, opposite (of), at, next to.	*Top, bottom, right, left, back, front.*

Table 39 A linguistic-based categorisation of elements carrying spatial meaning in spatial language[65] [66]

The spatial prepositions in the left column in Table 39 primarily more the spatial relation between two spatial objects, whereas spatial adverbs tend to highlight the location within the spatial object. The use of spatial adverbs seems to require the spatial object to have an upper, lower, back, front, right and left part, which requires a rather holistic approach to the spatial object (cf. Levinson, 1996; see Section 2.3.1.2). In contrast, spatial prepositions seem to reflect the analytic nature of representing spatial relations between two objects. These reflections lead to the formulation of hypotheses regarding their use in spatial discourse among the different sampling considered in this present study. Concerning the first possible influencing factor, language proficiency, no substantial difference is expected between the use of spatial adverbs and spatial prepositions, since both spatial adverbs and spatial prepositions are most likely to be acquired and used in everyday language. Hence, all students are expected to know prepositions and adverbs independent from their language proficiency, which leads to the formulation of hypothesis 7:

Hypothesis 7 (H7): No substantial difference is expected between the use of spatial prepositions and spatial adverbs among students with high and low language proficiency.

---

[65] Again, the data analysis was performed in the German language, hence the corresponding German spatial prepositions and adverbs have been used. However, in comparison to other languages, the German and English spatial prepositions and adverbs do not differ considerably in quantity and in meaning.

[66] The list of these elements in both categories is not complete, but rather based on the observed elements used by students in their spatial discourses in the solving of the spatial tasks.

With respect to students' spatial abilities, spatial language of students with high spatial abilities should show an increased spatial awareness, especially the greater use of spatial elements in their language to reflect their more developed spatial cognition (cf. Levinson, 2003; Coventry et al., 2009; Lehmann, 2013; see Section 2.3.1). Hence, spatial abilities should be an influencing factor on the structure of spatial language, as stated in the following hypothesis, hypothesis 8:

Hypothesis 8 (H8): Students with high spatial abilities are expected to make use of spatial prepositions and spatial adverbs more often in their spatial discourse than students with low spatial abilities.

The association of spatial adverbs with holistic spatial thinking and of spatial prepositions with analytic spatial thinking leads to the formation of another hypothesis concerning the structure of spatial language and the factor of sex. As indicated in previous studies (e.g., Rost, 1977), male students tend to show a preference for holistic approaches and better grasp of spatial orientation in spatial tasks and females show preference for analytic strategies, which seems to reflect the (analytic) predicative thinking, whereby the description of structural relationships are more in focus than concepts of spatial orientation (cf. Maier, 1999; Schwank, 2003). Hence, the following hypothesis can be formulated based on such observations:

Hypothesis 9 (H9): Female students are expected to use more spatial prepositions and less spatial adverbs than male students in their spatial discourse.

Let us take a look at the actual use of these linguistic elements in the describers' spatial language in the solving process in the reconstruction method under consideration of possible influencing factors. Figure 43 represents the results about the relative frequency of use of spatial prepositions and spatial adverbs in spatial discourse among the different four sample groups. The relative frequencies in Figure 43 represent the ratio of the spatial prepositions and spatial adverbs to the total number of phrases in the describers' spatial discourse.

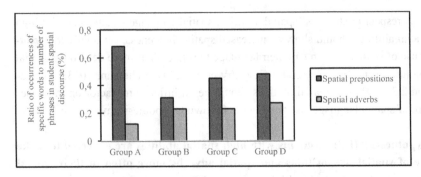

Figure 43 A linguistic-based analysis of students' use of spatial prepositions and spatial ad
verbs in their discourse among the four sampling groups

As it can be observed in Figure 43, students who achieved high scores in both
language proficiency and spatial ability tests seem to have used more spatial
prepositions and less spatial adverbs in their spatial discourse. For a deeper investigation of these observations, the data was analysed according to the different groups established by the separate dichotomies. For instance, a preliminary
analysis of students with high (Group A and Group B) and low language proficiency (Group C and Group D) showed that there is no substantial difference regarding the use of spatial prepositions and spatial adverbs, as visualised in Figure 44. However, further investigations show considerable difference between
students' use of spatial prepositions and adverbs under consideration of their
spatial abilities or sex, as visualised in Figure 45 and Figure 46 respectively.

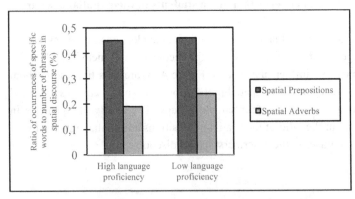

Figure 44 A linguistic-based analysis of students' use of spatial prepositions and adverbs in
their spatial discourse under consideration of their language proficiency

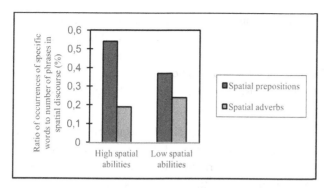

Figure 45 A linguistic-based analysis of students' use of spatial prepositions and adverbs in their discourse under consideration of their spatial abilties

In Figure 45, one can observe that students with high spatial abilities tend to use more spatial prepositions (i.e. 17 p.p. more) and slightly less spatial adverbs in their spatial discourse than students with low spatial abilities (i.e. 5 p.p. less). Overall, students with high spatial abilities seem to make more use of elements carrying spatial meaning in their discourse, more precisely 12 p.p., than students with low spatial abilities. Hence, this finding seems to support the statement formulated in Hypothesis 8, i.e. that students with higher spatial abilities tend to use more spatial adverbs and spatial prepostions, especially concerning the latter elements.

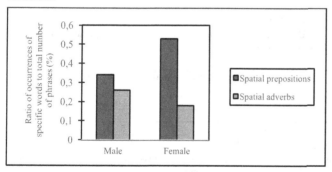

Figure 46 A linguistic-based analysis of students' use of spatial prepositions and adverbs in their discourse under consideration of their sex

Figure 46 shows the results for the linguistic-based analysis of spatial language concerning male and female participants. These results show that females tend to use more spatial prepositions (i.e. 19 p.p. more) and less spatial adverbs than their male counterparts in their spatial language (i.e. 8 p.p. less). Even though

199

these results are only based on preliminary observations of data distribution, these findings seem to be in line with the statement hypothesised in Hypothesis 9 (H9). In other words, female students tend to use more spatial prepositions for an analytic approach (the focus being spatial relations), and male students tend to use more spatial adverbs for a holistic approach (the focus being spatial orientation).

The trends observed in the above data using simple frequency methods were used for an analysis of any significance difference between the describers' use of particular linguistic means and the underlying background factors (especially, spatial abilities and gender) in this linguistic-based analysis approach to spatial language. Therefore, a statistical hypothesis test, the nonparametric Mann-Whitney U test, was applied to the collected data to investigate the above-mentioned hypotheses: H8 and H9. The following tables show selected results from the Mann-Whitney U tests for testing any significant difference between the variables sex and students' use of spatial prepositions or adverbs (see Table 40) and between spatial abilities and students' use of spatial prepositions or adverbs in spatial discourse (see Table 41):

	Spatial Prepositions	Spatial Adverbs
Mann-Whitney U	16.500	26.000
Wilcoxon W	52.500	62.000
Z	-1.631	-0.632
Asymptotic Significance (2-tailed)	0.103	0.527
Exact significance [2*(1-tailed Sig.]	0.105	0.574

Table 40   Mann-Whitney U test for testing the difference between use of spatial prepositions / adverbs in spatial discourse and sex

	Spatial Prepositions	Spatial Adverbs
Mann-Whitney U	24.500	26.000
Wilcoxon W	60.500	62.000
Z	-0.789	-0.632
Asymptotic Significance (2-tailed)	0.430	0.527
Exact significance [2*(1-tailed Sig.]	0.442	0.574

Table 41   Mann-Whitney U test for testing the difference between use of spatial prepositions / adverbs in spatial discourse and spatial abilities

The above findings show that even though preliminary results had shown that students' spatial abilities and sex may potentially exert influence on the structure of student's spatial language, no significant difference was observed in the statistical hypothesis tests (although a certain trend toward significance might be interpreted in the result achieved in Table 40, i.e. that spatial prepositions tended to be used more by a particular group, in this case, female participants). Hence, the results from the descriptive statistics show that students with higher spatial ability or female students do not significantly use more spatial prepositions in their spatial discourse, i.e. hypotheses H8 and H9 cannot be supported. With regards to the most promising influencing factor of spatial ability, these findings mean that the content analysis of spatial language from a linguistic-based approach does not show any significant difference between the structure of spatial language (regarding the linguistic means used) between students of high or low spatial ability. However, given that the students' intended meaning of these linguistic elements carrying spatial information was not deeply investigated in this approach and that spatial prepositions and adverbs can be used simultaneously in phrases of spatial language, another approach which takes into account the spatial meaning of linguistic elements needed to be considered. Such an approach, which is referred to as *content-based*, is introduced in the upcoming section.

6.2.3 Content-based approach

The content-based approach can be used for analysing student spatial language by categorising elements carrying spatial meaning based on the nature of spatial content, instead of according their linguistic property (as in the linguistic-based approach in Section 6.2.1). The need for such an alternative approach arises from the observations that some spatial prepositions or spatial adverbs seem to carry more specific spatial meaning in comparison to others. This leads to a new categorisation of elements carrying spatial meaning in spatial language in two categories – *relative* and *intrinsic* [67] - which are visualised in Table 42.

Relative elements	Intrinsic elements
Rechts, links, unter, vor, über, hinter, auf, oben, unten, hinten, vorne.	*An, bei, neben/seitlich, gegenüber, außerhalb, zwischen, mittig, in.*
(Right, left, under, in front of, over, behind of, on, at the top, at the bottom, at the back, at the front)	*(At, at, next to/at the side, across, outside, between, in the middle, in)*

Table 42   A content-based classification of elements carrying spatial meaning used in students' spatial discourse[68]

---

[67]   The meaning of the words relative and intrinsic slightly in this section differs from the meaning used in Levinson's (2003) reference frameworks for spatial orientation.

[68]   The list of these elements in both categories is not complete, but rather based on the observed elements used by students in their spatial discourses when solving the spatial tasks in the reconstruction method.

The categories *relative* and *intrinsic* in Table 42 reflect the intended spatial meaning of the elements carrying spatial information independent from their linguistic category. Please note that although the same names were used in previous literature, the meaning of these category names and their components differs in this chapter. Elements in the *relative* category consist of spatial prepositions and spatial adverbs which require the consideration of the subject's body for orientation and defining the back, front, left or right etc., which gives a more unambiguous meaning compared to elements in the intrinsic category. In the *instrinsic* category, the structure of the object as a ground is more in focus and the meaning of the elements is more ambiguous in space, whereby the subject's body is not explicitly emphasised in the elements used in spatial discourse (as in the defining the left, the right and front, and the back in the relative category). For instance, the decoding of the element *next to* can literally mean both *right* or *left* (and even *behind of* or *infront of*). Therefore, elements in the relative category seem to entail a more explicit unique spatial direction along the three dimensions in space, whereas elements in the instrinsic category lead to more interpretations of spatial knowledge in discourse; hence they are spatially more ambiguous than the former ones. Due to difference in spatial ambiguity between the categories, students with high spatial abilities are expected to use more elements which show spatial disambiguity in their discourse. Hence, relative elements should be used more by students with high spatial abilities, which can be formulated as a hypothesis for testing:

Hypothesis 10: Students with high spatial abilities are expected to use elements from the relative category more often, whereas students with low spatial abilities are expected to use more intrinstic elements in their spatial discourse.

Moreover, the fact that previous literature (e.g., Rost, 1977) suggested a better performance in spatial orientation for males leads to the preassumption that males should use more relative elements in their spatial discourse. The underlying hypothesis for the hypothesised effect of sex is:

Hypothesis 11: Male students are expected to use more relative and less intrinsic elements than female students in their spatial discourse.

Figure 47 represents the results of the use of spatial elements categorised according to the content-based approach among the four different sample groups. After breaking down the student spatial discourse in phrases, (average) ratios of the number of use of specific words – relative or instrinsic elements – to the (average) total number of phrases were developed for data relativisation concerning the length of student spatial discourse.

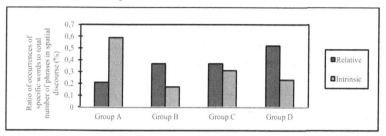

Figure 47   Representation of the use of relative and intrinstic spatial elements among the four sampling groups

One can notice that students in Group A used more spatial elements from the intrinsic category than students in the other three groups, and they have used the least number of spatial elements from the relative category from all groups. In contrast, participants of Group D have used the highest number of spatial elements from the relative category. This observation is rather unexpected, especially when considering that students with low spatial and low language proficiency (Group D) used 30 p.p. more spatially unambiguous elements in their spatial discourse than students with high achievements in both domains (see Figure 47). A more exact visualisation of the use of relative and intrinsic spatial elements between students with low and high spatial abilities is given in the following figure (see Figure 48).

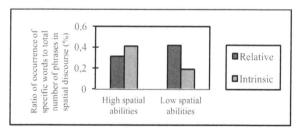

Figure 48   Use of relative and intrinsic spatial elements among students under consideration of their spatial abilities

The data in Figure 48 illustrates that students with high spatial abilities tended to use less relative elements and more intrinsic elements in their spatial discourse, when compared to students with low spatial abilities. For a deeper understanding of the results, let us take a look at specific relative and intrinsic elements which were used in spatial discourse of students with high spatial abilities compared to students with low spatial abilities (see Figure 49 and Figure 50). Figure 49 shows that former group made less use of elements, such as *left* and *right* (almost half as much students with low spatial abilities), which are important for spatial orientation. However, they made more use of spatial prepositions, such as *at* or *next to,* in their spatial language in contrast to students with low spatial abilities (see also Figure 50).

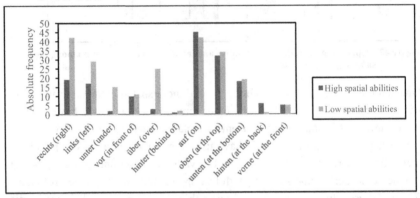

Figure 49 Distribution of use of relative spatial elements in students' spatial discourse under consideration of their spatial abilities

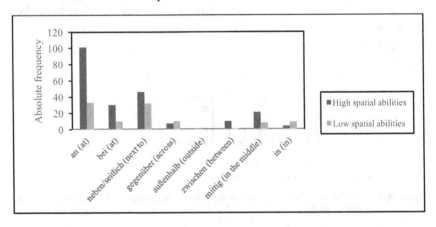

Figure 50 Distribution of use of intrinsic spatial elements in students' spatial discourse under consideratation of their spatial abilities

The apparent higher use of spatially ambiguous elements in spatial discourse of students with high spatial abilities, shows that hypothesis 10 cannot be validated on basis of these findings, and that their spatial language does not necessarily reflect a higher spatial awareness, which is expectable for students with more developed spatial abilities. This result was not anticipated, however, it is probable that the reason for this is the relatively higher reliance on spatial metaphors in their discourse (see results for Groups A and C in Table 32). Similar to these results for students sorted according to their spatial abilities, students with low language proficiency made more use of relative elements in their spatial discourse. However, another trend could be analysed when considering students' sex. As Figure 51 illustrates, the male participants used relatively more relative elements than intrinsic elements in their spatial language, whereas female students subtantially used more intrinsic elements.

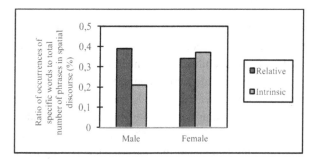

Figure 51 Use of relative and intrinsic spatial elements among students under consideration of their sex

Considering the fact, that relative spatial elements seem to be linked more to spatial orientation, whereby the spatial relations are required to be defined under consideration of the subject's body, the finding that males use relatively more spatial elements with less ambiguity could presumeably imply that males show a higher spatial awareness. It is important to note that generally female participants had longer spatial discourses when compared to male participants, which requires the results from such analyses to be treated with considerable caution. Nevertheless, the data in Figure 51 shows an indication to support the hypothesised effect of sex on structure of students' spatial knowledge in Hypothesis 11. For a statistical investigation of any significant difference concerning the formulated hypothesis H10, the data was tested using descriptive statistics, i.e. the nonparametric Mann-Whitney U test.

	Intrinsic	Relative
Mann-Whitney U	17.000	21.000
Wilcoxon W	53.000	57.000
Z	-1.580	-1.156
Asymptotic Significance (2-tailed)	.114	.248
Exact significance [2*(1-tailed Sig.]	.130	.279

Table 43   Mann-Whitney U test for testing the difference between use of intrinsic / relative elements in spatial discourse and spatial abilities

	Intrinsic	Relative
Mann-Whitney U	18.000	28.000
Wilcoxon W	54.000	64.000
Z	-1.475	-0.420
Asymptotic Significance (2-tailed)	.140	0.674
Exact significance [2*(1-tailed Sig.]	.161	0.721

Table 44   Mann-Whitney U test for testing the difference between use of intrinsic / relative elements in spatial discourse and sex

The results from these statistical descriptive tests (see Table 43 and Table 44) show that there is no significant difference between the use of intrinsic or relative elements and the variables of spatial abilities and gender. Hence, hypotheses 10 and 11 cannot be validated using the deductive approach in this study, in which data collected using qualitative research methods was additionally analysed using quantitative methods. The failure of showing significant difference in the underlying hypotheses raises further issues, as whether a qualitative data analysis is sufficient for investigating and supporting the results emerging from the simple frequency methods. Consider the following transcript excerpt from a female student with high spatial ability and high language proficiency, who has used the highest number of intrinsic elements in relation to the number of phrases in her spatial discourse:

Transcript 25:

(Describer 4 is describing spatial object A).

	(...)	*(...)*
Describer 4:	*Afterwards you have to uhm at one four so wa... that four, you must do at one so at it (...) So three, right? Next to each other. (...) And at the five at the foursome tower (...) you*	*Danach musst du ähm bei einem vier also wa... das ja vier musst du das halt bei einem dann so dranmachen (...) Also drei, ne? Nebeneinander halt. (...) Und bei den fünf bei so einem Viererturm (...) musst du halt*

must then... then stll a threesome tower next to it do it at it. (...) Then you must still do a double next to it in a way, so two cubes next to each other. (...) And then still a cube. Build [it] next to it.	*dann... dann noch so einen Dreierturm daneben dran tun. (...) Dann musst du noch einen Zweier daneben tun so, also zwei Steckwürfel daneben. (...) Und dann noch einen Steckwürfel. Daneben bauen.*

Builder 4: *Underneath?* *Unten?*

Describer 4: *Yes, underneath. Next to it.* *Ja, unten. Daneben halt.*

 (...) *(...)*

In Transcript 25, the spatial discourse of Describer 4 is characterised by the relatively high use of intrinsic elements (the ration of instrinsic elements to total number of phrases in her discourse is 0.75), which do not require a frame of reference of body orientation. The intrinsic elements *bei* (at), *an* (at), and *neben* (next to) increase ambiguity in spatial meaning and their high use in discourse does not necessarily reflect the high spatial abilities assessed according to the reference tests. Hence, Describer 4 did not necessarily seem to externalise the required spatial awareness in her spatial discourse in order to reduce spatial ambiguity, which raises questions about possible difference between assessment of spatial abilities: assessing spatial abilities using pencil-and-paper tests and using spatial discourse analysis. The latter type of assessment involves more complex phenomena, since the social-communicative skills play a more important role than in pencil-and-paper tests. Hence, an investigation of case studies on a qualitative method, such as the one above of Describer 4 in Transcript 25, shows that spatial discourse analysis for understanding or assessing spatial abilities are not a substitute for pencil-and-paper tests, but should rather be considered as an additional medium for a deeper understanding of student spatial thinking. A further case study in the following transcript, shows the high use of relative elements in the spatial discourse of Describer 14, a male student with low language proficiency and low spatial abilities (ratio of number of relative elements to number of phrases in his spatial discourse is 0.88).

Transcript 26:

(Describer 14 describes spatial object A after the control moment of the reconstruction method).

 (...) *(...)*

Describer 14:	Firstly, four stones in a row. On it, three in the right [one]... and the left must stay empty, the left stone. On it two, and the left one stays empty again. (...) And then one stone, and the left [one] stays empty again. Now built one at the topmost stone, at it, at the side. (...) Left. (...) Under it, under the left stone there are two again. Under it there are three under that which you have just built. (...)	Erstmal vier Steine in einer Reihe. (...). Darauf drei in die rechte... und das linke muss frei bleiben, der linke Stein. (...). Darauf zwei, und der linke bleibt wieder frei. (...) Und dann ein Stein, und das linke bleibt wieder frei. Jetzt baust du an dem obersten Stein eins dran an die Seite. (...) Links. (...) Darunter unter dem linken Stein kommen wieder zwei. Dadrunter kommen drei unter die du grad gebaut hast. (...)

In Transcript 26, Describer 14 used a high number of relative elements in his spatial discourse, e.g. *links* (left), *auf* (on), and *unter* (under), which reflect the use of body orientation and a reference frame for navigation in space. The high use of the relative elements might be triggered by the relatively less use of spatial strategies in contrast to other participants, which was identified and discussed in Section 6.1.5. Hence, a higher use of elements requiring body orientation and a reference frame in discourse when solving spatial tasks in the reconstruction method does not necessarily imply higher performance in spatial abilities tests, but can be used as an approach for avoiding development of specific strategies, e.g. breakdown strategy into two internal parts.

The above results show that structural analyses of spatial language using quantitative methods can offer deeper perspectives about different patterns observed among students under consideration of different variables (language proficiency, spatial abilities, and sex). However, such an approach is not enough for illustrating the complex interaction of the intendedness of use of particular elements in discourse. The above case studies in Transcript 25 and 26 emphasise that students, irrelative of their different background factors, exhibit different approaches for solving spatial tasks, which again influences the structure of the realised spatial language in discourse when solving spatial-verbal tasks. For instance, a higher use of specific elements in spatial discourse might be influenced by the use of particular strategies in the solving process.

6.2.4 Conception-based approach

In this section, I would like to investigate the nature of spatial language based on two conceptions which were introduced in the theoretical section (see Section 2.4.1). Whereas the structural and operational conceptions in Sfard's (1992) theoretical framework is based on mathematical concepts, I would like to extend the dual nature of spatial metaphors – static and dynamic – to the notion of spatial language as phrases which are either object-based or action-based. Object-based phrases of spatial language are characterised by reference to statements which describe the properties of spatial object from a static perspective. In contrast, action-based phrases are phrases which entail an action in space and are less object-oriented but rather action-oriented. Consider the following utterrances from two different describers in the reconstruction method:

| Describer 15: | *So at the bottom there are four cubes... four cubes next to each other, and at the right cube uhm there... at the right cube that one is empty, at the right-most* | *Also unten liegen vier Würfel... vier Würfel nebeneinander, und an dem rechten Würfel uhm da... an dem rechten Würfel der ist noch frei, an der ganz rechtem Würfel.* |
| Describer 4: | *You must click... four on top again so three or rather click on top... and then at the long, next to it, you must click one again to it, such one cube... and then build three high again.* | *Musst du auch nochmal vier oben draufsteck... also drei eher gesagt oben draufstecken... und bei den langen daneben musst du noch einen dranklicken so einen Würfel... und dann auch nochmal drei hochbauen.* |

The above utterrances in the spatial discourse of Describer 15 are characterised by the object-oriented description, which does not explicitly imply actions, but rather focuses on the properties of the spatial object. In contrast, the description by Describer 4 is action-based, whereby the describer focuses on the actions which need to be carried out in order to reconstruct the spatial object. In terms of Schwank's (2003) thinking styles, the approach adopted by Describer 15 can be allocated to the predicative thinking style, whereas the emphasis on actions by Describer 4 is typical for functional thinking. The influence of to which extent the spatial language is object-base or action-based should not be underestimated. Whereas the builder is required to reconstruct the actions for object construction by himself based on the static information in an object-based description, in the action-based phrases, the builder performs the given actions to perceive the spatial properties of the reconstructed object. The results of the analyses of the students' spatial language according to the conception-based approach are visualised in Figure 52, whereby all phrases in the spatial discourse of all describers (from Describer 1, D-1, to Describer 16, D-16) were classified as either object-based or action-based during data analyses.

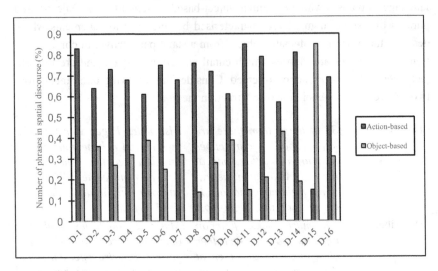

Figure 52 Dual analysis of the conception of the describers' spatial discourse

As Figure 52 illustrates, the majority of the students' spatial language seems to be action-based, which can be accounted for by the task instruction (see Section 4.2.2.3). No substiantial patterns could be observed in the results of spatial language analysis according to the dual conceptions under consideration of the students' language proficiency, spatial abilities, and sex (see Figure 52). However, curiously, one female student, D-15, shows a relatively much higher use of object-based spatial language in her discourse: 85% of the phrases in her spatial discourse were object-based and 15% of phrases were action-based (see Figure 52). The reason for this rather unexpected result is not entirely clear, however, given the relatively successful result in the solving of the spatial task A (description of spatial object A) (see solutions in Appendix B), this outlier might need further investigation from a qualitative point of view. In particular, such an investigation might address a possible relationship between highly object-based or action-based spatial language and task solutions. However, in the case of the present study, this additional focus would go beyond the scope of the underlying aims of this thesis.

6.3 Discussion of the results from the structural analyses of spatial language

In the above sections, I have investigated the structure of spatial language used by the students in solving the spatial-verbal tasks in the reconstruction method. I was interested in exploring any possible differences in the structure of spatial language used by students with different background factors. After having used spatial language as a medium for identifying and understanding spatial strategies and obstacles of students solving spatial tasks in the reconstruction method in Section 5.1 and Section 5.2 respectively, spatial language was required to be the object of analysis for exploring its nature among students of different language proficiency, spatial abilities, and sex. Different approaches to the analysis of spatial language were provided. Using a metaphoric-based approach, I have investigated to which extent student spatial language can be considered as metaphorical, by considering how often and which spatial metaphors are used for which purpose by students to structure their spatial thinking in discourse. The first finding showed that on average spatial metaphors are used in every third phrase in student's spatial discourse, regardless of the possible influencing background factors. Another preliminary finding illustrated that students with high spatial abilities seem to use spatial metaphors for describing the spatial positions of objects more often in their spatial discourse. However, due to the small sample no significant differences between use of specific type of metaphors and spatial abilities were observed. Overall, the findings show that most of the students, independent of possible influencing factors, tended to use more everyday spatial metaphors with structure function and of static nature. Unfortunately, the limited difference in values could not contribute for a deeper investigation of the relationship between students' language proficiency and use of spatial metaphors. Nevertheless, this approach revealed more about the nature of students' spatial language in solving verbal-spatial tasks, i.e. the metaphorical nature of spatial language of fifth grade students. The findings about the relatively high use of spatial metaphors shows how metaphors serve as a medium for spatial thinking among students. Spatial metaphors should not be merely considered as the externalisation of the underlying mental images generated during perception of spatial objects, but they also structure the space (re-)constructed in the spatial discourse. Hence, to a general extent, the finding implies that it be helpful for teachers or researchers to understand student's metaphorical spatial thinking, which can serve as a basis for further development of student's spatial language and ability. The strong evidence of metaphorical spatial thinking among students solving spatial-verbal tasks based on the findings about spatial metaphors and their use in Section 5.1 and Section 6.1 accordingly show the need for further development of existing models about spatial ability in mathematics education. Especially, spatial ability models emphasising language, e.g. the component of visual abilities in Pinkernell's model (2003) (see Section 2.1.3.4), can be extended by the metaphoric nature of spatial thinking concerning its representation in form of language.

The linguistic-based and content-based approach provide different categorisations of elements carrying spatial meaning used by students in their spatial discourse. In the former approach, a linguistic categorisation was applied on the elements carrying spatial meaning, spatial prepositions and spatial adverbs; excluding spatial metaphors and spatial deixis. As expected, the analysis did not reveal any substantial difference between use of spatial prepositions and spatial adverbs in spatial discourses of students with high or low language proficiency. However, the most striking result to emerge from the preliminary data analysis in this approach is that students with high spatial abilities use relatively more spatial prepositions and less spatial adverbs in their spatial discourse, compared to students with low spatial abilities. This seemed to reflect the higher emphasis on spatial relations among students with high spatial abilities, which might support the higher spatial awareness in discourse. Hence, under consideration of Levinson's (2003) assumption that an analysis of student's spatial language reflects the student's underlying spatial cognition, one might conclude that students with high spatial abilities are likely to think more in spatial relations than students with low spatial abilities. However, the results from the hypothesis tests in descriptive statistics have shown no significant difference between use of spatial prepositions or adverbs and the variable of spatial abilities.

Another preliminary finding indicated that female participants are also more likely to use spatial prepositions in their discourse than males. From a theoretical perspective, spatial prepositions seem to focus more on the analytic approach of

212

describing spatial relation between two objects, whereas spatial adverbs seem to reflect holistic approaches in spatial thinking. Hence, the above finding would substantiate previous findings in previous literature (e.g., Rost, 1977, Coluccia et al., 2007), which state that females are more likely to solve spatial tasks analytically and males use more holistic approaches in spatial tasks. However, no significant difference was observed between the use of specific elements in spatial discourse and sex.

In the content-based approach to spatial language, the same spatial-meaning carrying elements from the linguistic-based approach were categorised into two new categories: relative and the intrinctic elements. Relative elements are elements carrying spatial meaning which require more spatial orientation and reduce spatial ambiguity. In contrast, intrinsic elements are elements carrying spatial meaning which are solely based on the ground's structure and are highly ambiguous in their meaning. The most unexpected result to emerge from the data is that students with high spatial abilities tended to use more intrinsic elements in their spatial discourse, compared with students with low spatial abilities. It was expected that spatial discourse of students with high spatial abilities would be characterised by less spatially ambiguous elements, however, this was not the case. Regrettably, this present research has failed to give an entire plausible explanation for the low or high values of use of relative or intrinsic among students with high or low spatial abilities respectively. However, no significant difference between the use of relative/intrinsic elements in spatial discourse and student spatial abilities was identified. Further preliminary results about the influencing factor of sex and use of intrinsic and relative elements in spatial discourse seem to be in agreement with previous findings from literature (e.g., Rost, 1977). In particular, male students tended to use more relative elements demanding spatial orientation in their spatial discourse, when compared to female students. Overall, the linguistic-based and content-based approaches show the importance of considering both language and content based categorisation of elements carrying spatial meaning, each of which evoked different perspectives and expectations regarding the analysis of the same underlying data. This continues to strengthen the intertwining of language and spatial abilities which is emphasised throughout different phenomena in this present study. Even though the hypotheses formulated in above sections could not be validated in this present study, the explorative core of these structural approaches offered new perpectives in perceiving spatial abilities in discourse.

In the last approach to structural anaylsis of spatial language, the conception-based approach, spatial language was analysed in discourse according to its conceptual nature – either object-based or action-based. As expected, the findings show that students' spatial language tended to be highly action-based. However, the present study was not successful in giving an explanation for one outlier in Group D, whose spatial language was highly object-based in the discourse.

The above four approaches presented different ways of analysing student spatial language from different perspectives, which played an important role in characterising the nature of spatial language. Preliminary data analyses of student spatial language showed which kind of spatial metaphors are common in spatial discourse, but deeper investigations showed no significant difference between the use of metaphors in spatial discourse and the possible influencing factors of language proficiency, spatial abilities, and sex. Another approach to spatial language which revealed more about the nature of spatial language is the conception-based approach, which showed that spatial language of students solving the spatial tasks in the reconstruction method is highly action-based. The fact that no statistically significant results were determined in this present study, for instance, between students with low and high spatial abilities regarding their structure of spatial language (both in the linguistic-based and content-based approaches) gives rise to the question of to what extent spatial language can be indeed used to externalise spatial thinking, as an alternative to reference tests. Preliminary data analyses according to the content-based approach had even shown that students with spatial ability showed a higher tendency towards using intrinsic elements, which are more ambiguous in their meaning, in their spatial discourse. Such findings presumeably imply that while spatial language is an important medium for understanding how students think spatially, an analysis of distinct elements of spatial language is not sufficient to demonstrate the different student achievements in the reference tests for assessing spatial knowledge. Hence, spoken spatial language should not be considered as alternative for assessment of student spatial knowledge, but rather as an enriching tool for understanding and diagnosing student needs in spatial geometry.

7. Summary and Conclusion

In this final chapter, the findings provided in the previous chapter are discussed and synthesised for the summary of this thesis. Following the synthesis of the results, in Chapter 7.1, possible implications of the results for further research in mathematics education and for teaching mathematics are provided, followed by the concluding remarks in the last section of this chapter.

This present study focussed on how students solve spatial-verbal tasks in discourse. The first four research questions provided various findings concerning different aspects of dealing with the main research questions. The first research question (R1) dealt with possible strategies which students used when solving spatial-verbal tasks. In the second research question (R2), several obstacles which arised during the solving of the spatial-verbal tasks were described. The third and fourth research questions, (R3) and (R4), emphasised the role of possible influencing factors – students' language proficiency, spatial abilities, and sex – during the solving of the spatial tasks. While the third research question addressed whether the use of identified strategies varied with the influencing factors, the fourth one provided an insight into how spatial language can be analysed structurally using four different approaches and an explorative investigation on whether the use of spatial language varied with students' language proficiency, spatial abilities, or sex. An overview of the findings of all research questions is provided in the next section.

7.1 Synthesis of the results

The findings of this research study were described and discussed in Chapter 5 and Chapter 6. The detailed analysis of spatial language for identifying strategies and obstacles in solving spatial-verbal tasks in Chapter 5 provided an insight into understanding how students solve spatial-verbal tasks. In the results achieved in the inductive analyses in Chapter 5, spatial language served as a medium for reconstructing student spatial thinking in form of strategies or obstacles, which showed a strong empirical evidence of the relationship between spatial ability and language. In Chapter 6, the deductive analysis approach to the qualitatively collected data was triggered by the different background factors of students and their use of the identified strategies or the structure of their spatial language used in their spatial discourse for solving spatial tasks. In particular, spatial language served as an object of analysis in Chapter 6 and this explorative approach was intended to investigate whether the structure of spatial language in discourse differs among students with different background factors, e.g., students with high or low spatial abilities.

In the first part of Chapter 5, a range of strategies which students employed in the solving of the verbal-spatial tasks was described and allocated to analytic and holistic strategy groups (cf. Barratt, 1953). One of the most common strategies used (as observed in Section 6.1), spatial metaphors, showed how metaphors can serve as bridging tools for developing and communicating spatial knowledge between students. Due to the emphasis on particular characteristics or properties (of the object in source domain, which are then assigned to the spatial object), spatial metaphors were considered as a strategy group of an analytic nature. The high occurrences of spatial metaphor strategy in discourse enabled an in-depth classification of spatial metaphors, which was based on three dimensions: linguistic, spatial and conceptual. In the linguistic dimension, it was observed that different language means were used to externalise spatial metaphors – everyday, letter-based, or mathematical language. From a spatial content perspective, spatial metaphors were applied to communicate different spatial knowledge – structure, spatial position and spatial relations. In the conceptual dimension, the reconstruction of spatial knowledge in spatial metaphors was either of static or dynamic nature. The three dimensions of spatial metaphors led to the establishment of a model for spatial metaphors, which was illustrated in Section 5.1.1.4. The next two strategies, which were introduced in Section 5.1, were complementary strategies – the breakdown and assembling strategies – which addressed the way how students break down the spatial object and re-assemble the parts in which the object was broken into in their spatial discourse. These strategies emphasised primarily the mental manipulation of spatial objects or their internal parts, which classified them as holistic strategies. Another strategy of holistic nature, identified in the spatial discourse, was rotation strategy, whereby students described the rotation of an internal part of the object in their descriptive process of the spatial object. Other strategies included controlling strategies, cubes controlling and structure controlling strategies, in which describing students demanded feedback from the builder, regarding the number of cubes used or the structure of the reconstructed object respectively, in order to control their descriptive and interpretative process in the task. The different strategies show the diversity in stu-

216

dents' thinking and strategy development in the solving of spatial-verbal tasks, which were reconstructed using an analysis of students' spatial language. Moreover, the identification of such strategies in spatial discourse emphasise the need for an analysis of spatial language to investigate and understand student spatial thinking, in accordance with findings from Landau and Jackendoff (1993) and Levinson (1996, 2003), which would not have necessarily been identified using other methods, such as multiple-choice items in reference tests.

During the observations of students solving the designed spatial-verbal tasks, several student obstacles were identified, which were described in Section 5.2. Identified obstacles showed that not all identified strategies have had a beneficial effect in the student spatial discourse. For instance, the use of spatial metaphors is not always beneficial in solving spatial-verbal tasks, because they allow a high number of interpretations and hence are highly semantically ambiguous. This finding complies well with previous literature (e.g., Sfard, 1998; Malle, 2009) about the use of metaphors in mathematics learning and teaching. Another obstacle observed was the student's difficulty of verbalising spatial relations between objects, which seemed to be caused either by the lack of spatial awareness in their discourse or by the high cognitive demands of the underlying spatial task. The spatial tasks required the visual perception of spatial constellations, the processing of important spatial information, and their verbalisation. However, some students left out the verbalisation of necessary spatial information concerning the spatial objects, as it was described in Section 5.2.4. Other students showed difficulties in verbalising spatial manipulation, such as rotation of internal parts of the objects, in their description (see Section 5.2.3), which supports the previous findings about the increasing difficulties of verbalising dynamic spatial thinking (cf. Schwank, 2003; Plath, 2014). Taken together, the above obstacles show the need for support and fostering of spatial abilities and for further development of spatial language in geometry classroom: the underlying goal being improving students' spatial and language awareness while learning and developing spatial concepts.

Following the identification of strategies of fifth-grade students solving spatial tasks in the reconstruction method, the use of strategies was investigated for any dependency of possible influencing factors, which might play an important role in student strategy choice in the task solving process. Preliminary results showed that students with high language proficiency tend to use more identified strategies than students with low language proficiency. The overall higher use of identified strategies was also observed among students with high spatial abilities. Regarding the factor of sex, preliminary results illustrate that females tend to use slightly more spatial metaphors, assembling, rotation and structure controlling strategies compared to the male participants. Hypothesis testing was implemented for investigating significant differences between use of holistic or analytic strategies and the variables of language proficiency or spatial abilities or sex. However, no significant differences were observed between students' spatial ability and use of analytic or holistic strategies and between students' language proficiency and use of analytic or holistic strategies. In contrary to findings about female's higher use of analytic strategies and male's higher use of holistic strategies (cf. Coluccia et al., 2007), no significant differences regarding sex were determined between the use of analytic/holistic strategies in student spatial discourse. Additionally, it is important to note that these observations did not consider whether the use of these strategies was successful or not, hence the findings should be interpreted with caution. Moreover, these findings about the possible relationships between strategy choice and student's proficiency or spatial abilities or sex shall not be generalised due to the small sample size. However, I believe that this explorative approach served as a starting point for investigating these relationships and strengthening the assumption that strategy choice in spatial-verbal tasks does not only depend on the nature of the task, but also on the subject's underlying knowledge, which is in line with previous literature (cf. Rott, 2011; Plath 2014). Following the deductive analyses, it was agreed that additional qualitative investigations are required to observe how the subject's knowledge influences the strategy choice. This was illustrated in a case study in Section 6.1.5, which showed that a student with low language proficiency and low spatial abilities used a relatively low number of identified strategies by solely relying on spatial prepositions in his spatial description. According to Lemaire and Siegler (1995), the student's low use of different strategies and their flexible or adaptive use reflects low strategic abilities in solving spatial-verbal tasks, which might be influenced by his low language proficiency or low spatial ability or both. However, the present findings do not provide any statistically significant results regarding the use of strategies among students with high or low language proficiency and high or low spatial abilties.

The analysis of spatial language for determining strategies and obstacles also provided an insight in the different aspects of spatial language. These different quantified aspects are emphasised in the four different approaches for the structural analysis of spatial language, which were described in Sections 6.2.1, 6.2.2, 6.2.3 and 6.2.4. In particular, I was interested in whether the structure of student's spatial language may vary with the possible influencing factors of language proficiency, spatial abilities, and sex, under consideration of each structural approach. It was assumed that an analysis of spatial language can help understand student spatial thinking, and hence perhaps differences in student spatial abilities might also lead to a difference in the realisation of spatial language in discourse.

In the metaphorical approach to spatial language, I have investigated to what extent spatial language of fifth-grade students is of metaphorical nature. Preliminary results showed that students with high language proficiency tended to use more everyday, letter-based and mathematical spatial metaphors in their spatial discourse compared to students with low language proficiency (see Table 32). Further patterns emerging from preliminary data analysis include that female students tended to use more metaphors in their overall spatial discourse when compared to male participants and that students with high spatial abilities tended to describe more the spatial position of an object by using metaphors in their spatial discourse. However, no statistically significant differences between the use of specific types of spatial metaphors and variables, such as language proficiency, spatial abilities, and sex, were determined. Nevertheless, this analysis of spatial metaphors in student spatial discourse showed that the majority of metaphors used were of static nature independent from the students' language proficiency, spatial abilities, and sex. These findings showed that student spatial language is highly metaphorical, which can be triggered by the strong association to reality in students' spatial thinking. However, another cause for the high use of metaphors in spatial language can be the lack of linguistic means for verbalising spatial knowledge using spatial concepts, as it has been demonstrated and discussed in the students' obstacles. The results about spatial metaphors highlight the important role of metaphors in student spatial thinking when solving spatial-verbal tasks, which should be incorporated in existing spatial ability models in mathematics education (especially those models which consider the verbalisation of spatial knowledge and thinking as a spaital ability: e.g. Pinkernell, 2003). Additionally, the results about the diverse nature of students' spatial metaphors in Section 5.1 and about their high use in students' spatial discourse in Sections 6.1 and 6.2.1 suggest the need for addressing this phenomenon in learning of spatial geometry in order to support the transition from everyday language to mathematical language by referring to the relatively high degree of ambiguity in spatial meaning for raising spatial and language awareness (see Section 5.2.1). Spatial metaphors can act as a support in introduction of geometrical and spatial concepts, but their broad meaning can be used to show the necessity for students to develop mathematics language.

Apart from spatial metaphors, spatial adverbs and prepositions are other linguistic means for verbalising and communicating spatial knowledge and thinking. In the linguistic approach to spatial language, it was investigated whether the use of spatial adverbs and prepositions varies among students with different background factors. Preliminary findings showed that spatial discourse of students with high spatial abilities seems to be structured more by the use of spatial prepositions, which might reflect the higher spatial awareness in verbalising spatial relations between spatial objects. However, no statistically significant difference between the use of spatial prepositions or adverbs and student spatial abilities was observed in this study.

Further observations from preliminary data analyses include that female participants tended to use more spatial prepositions and less spatial adverbs in their spatial discourse compared to their male counterparts. However, no statistically significant difference between the use of spatial prepositions or adverbs and sex was determined. Apart from the linguistic classification of the elements carrying spatial meaning in spatial language, a more spatial-oriented classification, the content-based approach, was introduced for analysing student spatial language in discourse. In this approach, the spatial prepositions and adverbs were recategorised in two groups: relative (spatially less ambiguous) and intrinsic elements (spatially more ambiguous). Contrary to expectations, preliminary findings showed that students with high spatial abilities use more intrinsic elements in their spatial discourse than students with low spatial abilities. The reason for this rather contradictory result is not entirely clear, but this finding might be linked to the higher use of spatial metaphors among students with high spatial abilities. Hence, spatial metaphors could have compensated for the lower use of relative elements in student spatial discourse. However, the preliminary results should be treated with caution, because no statistically significant difference was observed between the use of intrinsic or relative elements and the variables of spatial abilities or sex. Regarding the sex factor, preliminary trends in data analyses show that males use less intrinsic elements (elements which do not necessarily require spatial orientation) in their spatial discourse than females. This supports the assumption from previous literature (e.g., Rost, 1977) that males are better at orientating themselves in space in contrast to females, and that an analysis of spatial language might show the sex differences in solving of spatial tasks. However, even here no statistically significant difference was determined between the use of intrinsic or relative elements and sex.

In the last approach to spatial language, the conception-based approach, spatial language was analysed based on the conceptual nature of the underlying phrases in the spatial discourse. The findings in Section 6.2.4 show that almost all students used more action-based language in their spatial description, which might be highly dependent from the nature of the task instruction. Only one student used a predominantly object-based spatial language, which raised questions with regard to the influence of the conceptions of spatial language on the solving process, given that this particular student managed to solve the first spatial task successfully (i.e. spatial object A was rebuilt almost identically by the builder; see Appendix B). However, due to the high influence of the builder's interpretations of the descriptions and the underlying actions on the results, i.e. the structure of the rebuilt object at the end of each task, the present study did not investigate solely the end results, but rather focussed on the processes taking place during the task solving process.

These four approaches to a structural analysis of spatial language should not be regarded as clear-cut distinct, but as ways of describing the different facets of the notion of spatial language. Spatial language should not be merely considered as the encoding of object properties or solely as the language of objects and places – as described by Landau and Jackendoff (1993) – however, it also involves the externalisation of dynamic spatial manipulation and spatial images which are mediated via spatial metaphors. Moreover, findings of this present study support previous literature on spatial language (e.g., Landau & Jackendoff, 1993; Coventry et al., 2009) stating that spatial language is an important tool for understanding student spatial thinking in solving spatial tasks, which was manifested in the description of strategies developed by and obstacles encountered by students solving spatial-verbal tasks. In comparison to pencil-and-paper tests, an analysis of spatial language in discourse enables researchers to understand strategies and obstacles, however, results from this present study do not provide an alternative form of assessment for spatial abilities. Deductive analyses of student spatial language showed no significant differences regarding the structure of spatial language between students with high spatial abilities and students with low spatial abilities. However, such statements should be investigated in a feasibile study with a bigger sampling group in future research in mathematics education.

Regarding the methodological aspect of this present study, the methodological framework presented in Chapter 3 provided additional evidence for the efficacy of the reconstruction method as a distinctive qualitative research method, especially when considering the research of spatial language and content learning. The observations and theoretical foundations of the reconstruction method strengthen and justify the adequacy and efficacy for developing and implementing spatial tasks which embody the notion of spatial abilities in mathematics education (e.g., Pinkernell, 2003), in which the describing, understanding, interpreting, modelling of spatial phenomena play an important role.

The results of this study show how interweaved language and content learning can be. The interplay between language and spatial content is reflected in the findings of this present study; in particular, in the verbalisation of spatial concepts and thinking and their interpretation in discourse, in the possible influence of the subject's language proficiency and spatial abilities for developing strategies, and in the ways of analysing spatial language structurally. Taken together, the above findings emphasise the importance of analysis of spatial language as a powerful tool for understanding students' spatial thinking and ways of approaching spatial tasks. In particular, the results show the need for supporting the development of spatial language and spatial concepts in geometry lessons. In order to foster spatial and language learning in mathematics classroom, teachers and teacher educators must understand the nature of student's spatial language, which was intensively investigated and promoted in this present study.

7.2 Implications

In this section, some possible implications of this present study for future research are illustrated. To begin with, due to the present study being small scale, it might be worthwhile to investigate certain aspects from a larger scale perspective involving 'pure' quantitative methods, which might imply the consideration of a larger sampling group and students of different ages, especially for a deeper understanding of the phenomena induced by the third and fourth research questions involving deductive analyses. In particular, further research shall be carried out to investigate the relationship between students' results in mental rotation tests and the use of rotation strategy in spatial-verbal tasks and the investigation between task solutions and possible influencing factors – language proficiency, spatial abilities, and sex – on a large scale, as indicated in previous literature (e.g., Linn & Petersen, 1985). A larger sampling group enables a more profound research regarding the choice of strategies among students solving spatial tasks for testing formulated hypotheses based on previous literature (e.g., Geiser et al., 2006; Coluccia et al., 2007). For a better generalisation and analysis of spatial language, future work should focus on the development of spatial language among students of different ages. This present study has only given an insight in spatial language in discourse of fifth-grade students, based on the assumption that students at this level encounter more obstacles in solving spatial-verbal tasks. However, future research can investigate the use of spatial language among students in more advanced classes, which could lead to new perpectives or approaches to the notion and understanding of spatial language.

Concerning the research method applied in this present study, I suggest further research on the use and development of this research method for investigating language and content learning in the reconstruction method. Further research could profit from the theoretical foundation of the reconstruction method presented in this present study and use it as a diagnostic and language support tool for assessing students' linguistic proficiency and concept images and by additionally providing language support by means of micro-scaffolding (see Section 3.2.2). From this perspective, future research about language and content learning in spatial geometry can establish which linguistic tools and concepts need to be strategically promoted, and how the students react, understand, learn and apply scaffolds during the solving of spatial-verbal tasks.

Moreover, it is plausible that a number of limitations could have influenced the results obtained. To begin with the limited number of spatial objects used in the main study. Further experimental investigations are needed to establish whether other different object structures might enable the identification of further strategies or obstacles, which might have not been accounted for in the present findings.

An implication for teaching which can be derived from the findings of this present study is the need for further emphasis of development and fostering of spatial language in mathematics classrooms and curriculum development in mathematics education. Spatial concepts and their verbalisations are only addressed to a limited extent in current mathematics lessons, however, as findings about the obstacles encountered by students solving the spatial-verbal tasks show, students' underdeveloped spatial language and other spatial abilities require support and further development, which can only be trained in geometry classes. Hence, spatial abilities and the underlying spatial language should be less considered as a prequisite for successful learning of geometry and other topics, but rather explicitly trained and developed in mathematics classrooms. Although the tasks designed for this study are rather specific, the majority of the obstacles identified in Chapter 5.2 can be justified by a lack of training in verbalising spatial and geometrical issues, which is not documented enough in the section of figures and space in the German mathematics curriculum at the beginning of secondary school. Such competencies can be trained in geometry classes by involving students to talk about, discuss, and describe spatial objects to others, which can be implemented in a language and content integrated learning setting in mathematics classroom.

Another implication for the teaching of spatial geometry is the importance of use of spatial metaphors as scaffolds not only for building upon the existing student knowledge, but also for showing the limitations of spatial metaphors for enabling an exact and unambiguous meaning. Spatial metaphors can be used for introducing spatial-geometrical concepts based on everyday knowledge, and by highlighting their limitations for a generalised or highly abstact definition of the same concept in mathematics. Hence, the explicit discussion about spatial metaphors in classrooms would not only increase students' spatial and language awareness, but also triggers students' understanding about the importance of learning and developing mathematics language and its use in mathematical discourse.

7.3 Concluding remarks

This present study provides insights into how students solve spatial-verbal tasks by focussing on their spatial language for reconstructing their underlying spatial knowledge. By identifying strategies used by students to solve verbal-spatial tasks, the thesis offers a wide range of approaches developed by the students in discourse in order to reach the underlying goals (see Chapter 5.1). Obstacles primarily related to the verbalisation of spatial knowledge or thinking were illustrated and discussed in qualitative analyses discussed in Chapter 5.2. Moroever, the use of strategies was investigated in relation to factors – students' language proficiency, spatial abilities, and sex – which could have had an influence on the task solving process (see Chapter 6.1). Rather than only considering spatial language as a medium for reconstructing conceptual and heuristic knowledge, in Chapter 6.2, spatial language served as an object of analysis, whereby different approaches for structural analysis of students' spatial language were considered. An analysis of elements of student spatial language enabled the investigation of the nature of student spatial language during the solving of spatial tasks in the reconstruction method. Findings showed the highly metaphorical and action-based nature of student spatial language. However, results from the deductive analyses did not show any statistically significant difference differs among students with different background factors (e.g., students with low or high spatial abilities) regarding the structure of their spatial language. Such findings emphasise that whilst spatial language is an important tool for understanding student spatial thinking, an analysis of spatial language based on the approaches introduced in this thesis does not seem to be an adeuqate tool for assessing student performance in spatial abilities.

Although the present study is of a rather qualitative nature, the diversity represented in the sampling enabled data analysis from a mixed methodological perspective. Such explorative findings shall not be conceived as an attempt to give a global idea of spatial language in all spatial tasks, but they should be perceived as in-depth observations of specific spatial tasks, which contribute for an indepth analysis of the notion of students' spatial language and of how students solve such spatial-verbal tasks. I believe that the investigation of a phenomenon from a specialised perspective can contribute for a better understanding of the generalised underlying phenomenon. In the case of this study, the investigation of spatial language in a particular context shall shed light on developing a definition and to a certain extent delineate it from other varieties of language in mathematics education, in order to provide a framework for generalising this concept in mathematics education – which is required for providing support in language and content learning in spatial geometry. Teachers and teacher educators require a better understanding of students' language and concept image, which can be reconstructed by an analysis of language, to improve their teaching style and supporting children in their language and content learning in mathematics classes by building up on the existent linguistic and conceptual knowledge. Hence, teacher educations, teachers and prospective teachers are possible target groups benefitting from a study of the findings of this research.

References

Abedi, J., & Lord, C. (2001). The language factor in mathematics tests. *Applied Measurement in Education, 14*, 219–234.

Aebli, H. (1980). *Denken: das Ordnen des Tuns. Vol. I: Kognitive Aspekte der Handlungstheorie* [Thinking: the organisation of doing. Vol. I: Cognitive aspects of the action theory]. Stuttgart: Klett-Cotta.

Baker, C. (2006). *Foundations of bilingual education and bilingualism* (4th Edition). Buffalo, Multilingual Matters.

Barratt, E. S. (1953). An analysis of verbal reports of solving spatial problems as aid in defining spatial factors. *Journal of Psychology, 36*, 17–25.

Barwell, R. (2009). *Multilingualism in mathematics classrooms. Global perspectives*. Bristol: Multilingual Matters.

Bennett, G. K., Seashore, H. G., & Wesman, A. G. (1973). *Differential Aptitude Tests (DAT)*. New York: Psychological Corporation.

Berelson, B. (1952). *Content analysis in communication research*. Glencoe: Free Press.

Bergvall, V. L., Sorby, S. A., & Worthen, J. B. (1994). Thawing the freezing climate for women in engineering education: Views from both sides of the desk. *Journal of Women and Minorities in Science and Engineering, 1*, 323–346.

Birkel, P. C., Schein, S. A., & Schumann., H. (2002). *Bausteine-Test. Ein Test zur Erfassung des räumlichen Vorstellungsvermögen* [Building cubes test. A test for assessing spatial abilities] (1st Edition). Göttingen: Hogrefe.

Büchter, A. (2011). *Zur Erforschung von Mathematikleistung. Theoretische Studie und empirische Untersuchung des Einflussfaktors Raumvorstellung* [About the research on mathematics performance. A theoretical study and empirical investigation of the influencing factor of spatial ability] (Doctoral Dissertation). Dortmund: Technische Universität Dortmund.

Bruner, J. S. (1971). *Toward a theory of instruction*. Cambridge: Harvard University Press.

Bruner, J. S., Olver, R. R., & Greenfield, P. M. (1988). *Studien zur kognitiven Entwicklung* [Studies about cognitive development]. Stuttgart: Klett.

Burin, D. I., Delgado, A. R., & Prieto, G. (2000). Solution strategies and gender differences in spatial visualization tasks. *Psicológica, 21,* 275–286.

Burton, L. A., Henninger, D., & Hafetz, J. (2005). Gender differences in relations of mental rotation, verbal fluency, and SAT scores to finger length ratios as hormonal indexes. *Developmental Neuropsychology, 28*(1), 493–505.

Caldera, Y. M., Culp, A. M., O'Brien, M., Truglio, R. T., Alvarez, M., & Huston, A. C. (1999). Children's play preferences, construction play with blocks and visual-spatial Skills: are they related? *International Journal of Behavioural Development, 23*(4), 855–872.

Carpenter, T. P. (1988). Teaching as problem solving. In: E. A. Silver (Ed.), *The teaching and assessing of mathematical problem solving* (pp. 187–202). Hillsdale: Erlbaum.

Carr, W., & Kemmis, S. (1986). *Becoming Critical. Education, Knowledge and Action Research.* London: RoutledgeFarmer.

Carroll, J. B. (1988). *Language, thought, and reality: selected writings of Benjamin Lee Whorf.* Cambridge: MIT Press.

Casey, B. M., Andrews, N., Schindler, H., Kersh, J. E., Samper, A., & Copley, J. (2008). The development of spatial skills through interventions involving block building activities. *Cognition and Instruction, 26,* 269–309.

Charters, E. (2003). The use of think-aloud methods in qualitative research. An introduction to think-aloud methods. *Brock Education, 12*(2), 68–82.

Chomsky, N. (1995). *The Minimalist Program.* Cambridge: MIT Press.

Clarkson, P. C. (1992). Language and mathematics: a comparison of bilingual and monolingual students of mathematics. *Educational Studies in Mathematics 23*(4), 417–429.

Clarkson, P. C. (2009). Potential lessons for teaching in multilingual mathematics classrooms in Australia and Southeast Asia. *Journal of Science and Mathematics Education in Southeast Asia, 32*(1), 1–17.

Clements, M. A. (1983). The question of how spatial ability is defined, and its relevance to mathematics education. *Zentralblatt für Didaktik der Mathematik, 15,* 8–20.

Cohen, L., & Manion, L. (1994). *Research methods in education* (4th Edition). London: Routledge.

Coluccia, E., Iosue, G., & Bradimonte, M. A. (2007). The relationship between map drawing and spatial orientation abilities: A study of gender differences. *Journal of Environmental Psychology, 27*(2), 135–144.

Coventry, K. R., Tenbrink, T., & Bateman, J. (2009). *Spatial language and dialogue.* Oxford: Oxford University Press.

Creswell, J. W. (2003). *Research design. qualitative, quantitative, and mixed methods approaches* (2nd Edition). London: Sage.

Cummins, J. (1979). Linguistic interdependence and the educational development of bilingual children. *Review of Educational Research 49,* 222–251.

Denzin, N. K., & Lincoln, Y. S. (2005). Introduction: The discipline and practice of qualitative research. In: N. K. Denzin & Y. S. Lincoln (Eds.), *The sage handbook of qualitative research* (2nd Edition). Thousand Oaks: Sage.

Diessel, H., & Hilpert, M. (2016). Frequency Effects in Grammar. In: M. Aronoff (Ed.), *Oxford Research Encyclopedia of Linguistics.* New York: Oxford University Press.

Durkin, K. (1991). Language and mathematical education: An introduction. In: K. Durkin & B. Shire (Eds.), *Language in Mathematical Education: Research and Practice* (pp. 3 –16). Milton Keynes, Philadelphia: Open University.

Ericsson, K. A., & Simon, H. A. (1980). Verbal reports as data. *Psychological Review, 87*(3), 215–251.

Evans, N., & Levinson, S. C. (2009). The myth of language universals: Language diversity and its importance for cognitive science. *Behavioral and Brain Sciences, 32,* 429–492.

Fagot, B., & Littman, I. (1976). Relation of preschool sex-typing to intellectual performance in elementary school. *Psychological Reports, 39,* 699–704.

Farias, M. (2005). Critical language awareness in foreign language learning. *Literatura y Lingüítica, 16,* 211–222.

Federal German Ministry of Education. (2004). *Beschlüsse der Kultusministerkonferenz. Bildungsstandards im Fach Mathematik für den Mittleren Schulabschluss* [Decisions of the ministry of education and cultural affairs. Educational standards for the intermediate school leaving certificate in mathematics]. Darmstadt: Betz. Retrieved from http://www.kmk.org/fileadmin/Dateien/veroeffentlichungen_beschluesse/200 3/2003_12_04-Bildungsstandards-Mathe-Mittleren-SA.pdf

Fennema, E. (1974). Mathematics learning and the sexes: A review. *Journal for Research in Mathematics Education, 5,* 126–139.

Finke, R. A. (1989). *Principles of mental imagery.* Cambridge: MIT Press.

Flick, U., von Kardorff, E., & Steinke, I. (2004). What is qualitative research? In: U. Flick, E. von Kardoff, & I. Steinke (Eds.), *A Companion to Qualitative Research.* (pp. 3–11). London: Sage Publications.

Font, V., Godino, J., Planas, N., & Acevedo, J. I. (2010). The object metaphor and sinecdoque in mathematics classroom discourse. *For the Learning of Mathematics, 30,* 15–19.

French, J., Ekstrom, R., & Price, L. (1963). *Kit of reference tests for cognitive factors.* Princeton: Educational Testing Service.

Freud, S. (1982). Angst und Triebleben (32. Vorlesung) [Fear and human drive (32th lecture)]. In: S. Freud (Ed.), *Vorlesungen zur Einführung in die Psychoanalyse und Neue Folge* (Vol. 1). Frankfurt am Main: Fischer.

Friedrich, H. F., & Mandl, H. (2006). Lernstrategien: Zur Strukturierung des Forschungsfeldes [Learning strategies: on the structuring of the research field]. In: H. Mandl & H. F. Friedrich (Eds.), *Handbuch Lernstrategien* (pp. 1–23). Göttingen: Hogrefe.

Frostig, M., Horne, D., & Miller, A. (1977). *Visuelle Wahrnehmungsförderung: Übungs- und Beobachtungsfolge für den Elementar- und Primarbereich* (2nd

Edition) [Visual perception: exercise and study for the elementary and primary level]. Dortmund: Crüwell.

Fülöp, E. (2015). Teaching problem-solving strategies in mathematics. *Lumat, 3*(1), 37–54.

Gardner, H. (2006). *Multiple Intelligences. New Horizons.* New York: Basic Books.

Geiser, C., Lehmann, W., & Eid, M. (2006). Separating "Rotators" From "Nonrotators" in the Mental Rotations Test. A Multigroup Latent Class Analysis. *Multivariate Behavioural Research, 41*(3), 261–293.

Gibbons, P. (2006). *Bridging discourses in the ESL classroom. Students, teachers and researchers.* London: Continuum.

Gittler, G. (1984). Entwicklung und Erprobung eines neuen Testinstruments zur Messung des räumlichen Vorstellungsvermögens [Development and implementation of a new test instrument for measuring spatial ability]. *Zeitschrift für Differentielle und Diagnostische Psychologie, 2*, 141–165.

Glaser, B. (1978). *Theoretical sensitivity. Advances in the methodology of grounded theory.* San Francisco: The Sociology Press.

Glaser, B., & Strauss, A. (1967). *The discovery of grounded theory: strategies for qualitative research.* Chicago: Aldine.

Grüßing, M. (2002). Wieviel Raumvorstellung braucht man für Raumvorstellungsaufgaben? Strategien von Grundschulkindern bei der Bewältigung räumlich-geometrischer Anforderungen [How much spatial ability does one require for spatial abilty tasks? Strategies of primary school students during solving of spatial-geometrical tasks]. *Zentralblatt für Didaktik der Mathematik, 34*(2), 37–45.

Guba, E. G., & Lincoln, Y. S. (2005). Paradigmatic controversies, contradictions, and emerging confluences. In: N. K. Denzin & Y. S. Lincoln (Eds.), *The Sage handbook of qualitative research* (3rd Edition, pp. 191–215). London: Sage.

Guay, R. B., & McDaniel, E. D. (1977). The relationship between mathematics achievement and spatial abilities among elementary school children. *Journal for Research in Mathematics Education, 8*(3), 211–215.

Hassanzadeh, N., Shayegh, K., & Hoseini, F. (2011). The impact of education and awareness in mother tongue grammar on learning foreign language writing skill. *Journal of Academic and Applied Studies, 1*(3), 39–59.

Ilgner, K. (1974). Die Entwicklung des räumlichen Vorstellungsvermögens von Klasse 1 bis 10 [The development of spatial ability from class 1 to 10]. *Mathematik in der Schule, 12,* 693–714.

Jordens, P., & Lalleman, J. A. (1988). *Language development.* Dordrecht: Foris.

Jungwirth, H. (2003). Interpretative Forschung in der Mathematikdidaktik. Ein Überblick für Irrgäste, Teilzieher und Strandvögel [Interpretative research in mathematics education. An overview for vagrants, partial migrants and resident birds]. *Zentralblatt für Didaktik der Mathematik, 35*(5), 189–200.

Kantowski, M. G. (1980). Some thoughts on teaching for problem solving. In: S. Krulik & R. Reys (Eds.), *Problem solving in school mathematics: 1980 yearbook* (pp. 195–203). Reston: National Council of Teachers of Mathematics.

Kelle, U., & Kluge, Susann. (2010). *Vom Einzelfall zum Typus. Fallvergleich und Fallkontrastierung in der qualitativen Sozialforschung* [From the individual case to the type. Case comparison and case contrast in the qualitative social research] (2th Edition). Wiesbaden: Springer.

Klein-Barley, C., & Raatz, U. (1984). A survey of research on the C-Test. *Language Testing, 1,* 134–146.

Knapp, W. (2006). Sprachunterricht als Unterrichtsprinzip und Unterrichtsfach [Language lessons as teaching principle and subject]. In: U. Brendel, H. Günther, P. Klotz, J. Ossner & G. Siebert-Ott (Eds.), *Didaktik der deutschen Sprache* (pp. 589–601). Paderborn: Ferdinand Schöningh.

Krause, C. M. (2016). *The mathematics in our hands. How gestures contribute to constructing mathematical knowledge.* Heidelberg: Springer Spektrum.

Kuhl, J., & Beckmann, J. (1985). Historical perspectives in the study of action control. In: J. Kuhl & J. Beckmann (Eds.), *Action control: From cognition to behaviour* (pp. 89–100). Heidelberg, New York: Springer.

Lakoff, G., & Johnson, M. (1980). The metaphorical structure of the human conceptual system. *Cognitive Science, 4*, 195–208.

Lakoff, G., & Núñez, R. E. (2000). *How the embodied mind brings mathematics into being.* New York: Basic Books.

Lanca, M., & Kirby, J. R. (1995). The benefits of verbal and spatial tasks in contour map learning. *Catrographic Perspectives, 21*, 3–15.

Landau, B., & Jackendoff, R. (1993). "What" and "where" in spatial language and spatial cognition. *Behavioral and Brain Sciences, 16*, 217–265.

Langacker, R. W. (1987). *Foundations of Cognitive Grammar, Volume 1.* Stanford: Stanford University Press.

Lehmann, C. (2013, September 19). *Raumorientierung: Kognition und Sprache* [Spatial orientation: Cognition and Language]. Retrieved from http://www.christianlehmann.eu/

Leisen, J. (2011). Sprachsensibler Fachunterricht. Ein Ansatz zur Sprachförderung im mathematisch-naturwissenschaftlichen Unterricht [Language-sensible subject learning. An approach for language support in mathematics and science classes]. In: S. Prediger, & E. Özdil (Eds.), *Mathematiklernen unter Bedingungen der Mehrsprachigkeit* (pp. 143–162). Münster: Waxmann.

Lemaire, P., & Siegler, R. S. (1995). Four aspects of strategic change. Contributions to children's learning of multiplication. *Journal of Experimental Psychology, 124*, 83–97.

Lengnink, K., Meyer M., & Siebel, F. (2014). MAT(H)Erial. *Praxis der Mathematik in der Schule, 58*(14), 1–10.

Lesser, R. (1989). Language in the brain: Neurolinguistics. In: N. E. Collinge (Ed.), *An Encyclopedia of Language* (pp. 205–231). New York: Routledge.

Levinson, S. C. (1994). Immanuel Kant among the Tenejapans; anthropology as empirical philosophy. *Ethos, 22*(1), 3–41.

Levinson, S. C. (1996). Language and space. *Annual Review of Anthropology, 25*, 353–382.

Levinson, S. C. (2003). *Space in language and cognition: Explorations in cognitive diversity.* Cambridge: Cambridge University Press.

Linn, M. C., & Petersen, A. C. (1985). Emergence and characterization of sex differences in spatial ability: a meta-analysis. *Child Development, 56,* 1479–1498.

Logan, G. D. (1988). Towards an instance theory of automatization. *Psychological Review, 95,* 492–527.

Lurija, A. R. (1992). *Das Gehirn in Aktion. Einführung in die Neuropsychologie* [The brain in action. An introduction to neuropsychology]. Reinbek: Rowohlt.

Lyons, J. (1981). *Language and linguistics.* Cambridge: Cambridge University Press.

Maier, P.-H. (1999). *Räumliches Vorstellungsvermögen. Ein theoretischer Abriß des Phänomens räumliches Vorstellungsvermögen* [Spatial ability. A theoretical outline of the phenomenon of spatial ability]. Donauwörth: Auer.

Maier, H., & Schweiger, F. (1999). *Mathematik und Sprache* [Mathematics and language]. Vienna: Öbv & hpt.

Malle, G. (2009). Mathematiker reden in Metaphern [Mathematicians speak in metaphors]. *Mathematik lehren, 156,* 10–15.

Masters, M. S., & Sanders, B. (1993). Is the gender difference in mental rotation disappearing? *Behaviour Genetics, 23*(4), 337–341.

Mayring, P. (2015). Qualitative content analysis. Theoretical background and procedures. In: A. Bikner-Ahsbahs, C. Knipping & N. Presmeg (Eds.), *Approaches to Qualitative Research in Mathematics Education. Examples of Methodology and Methods* (pp. 365–380). Heidelberg: Springer.

Mertens, D. M. (2015). *Research methods in education and psychology. Integrating diversity with quantitative and qualitative approaches* (2nd Edition). London: Sage.

Michael, W. B., Guilford, J. P., Fruchter, B., & Zimmermann, W. S. (1957). The description of spatial-visualization abilities. *Educational and Psychological Measurement, 17,* 185–199.

Morgan, C. (2004). Word, definitions and concepts in discourses of mathematics, teaching and learning. *Language and Education, 18,* 1–15.

Moschkovich, J. N. (1999). Supporting the participation of English language learners in mathematical discussions. *For the Learning of Mathematics, 19*(1), 11–19.

Moschkovich, J. N. (2010). *Language and mathematics education. Multiple perspectives and directions for research.* Charlotte: Information Age Publishing.

Noddings, N. (1990). Construtivism in Mathematics Education. *Journal for Research in Mathematics Education, 4,* 7–18.

North-Rhine Westphalia Ministry of Education. (2011). *Kernlehrplan und Richtlinien für die Hauptschule in Nordrhein-Westfalen* [Curriculum for the secondary schools in North-Rhine Westphalia]. Retrieved from http://www.schulentwicklung.nrw.de/lehrplaene/upload/lehrplaene_downloa d/hauptschule/Mathe_HS_KLP_Endfassung.pdf

North-Rhine Westphalia Ministry of Education. (2014). Kernlehrplan für die Sekundarstufe II Gymnasium/Gesamtschule in Nordrhein-Westfalen [Curriculum for the upper secondary schools Gymnasium/Gesamtschulen in North-Rhine Westphalia]. Retrieved from http://www.schulent- wicklung.nrw.de/ lehrplaene/upload/klp_SII/m/KLP_GOSt_Mathematik.pdf

Núñez, R. (2000). Mathematical Idea Analysis: What Embodied Cognitive Science can say about the Human Nature of Mathematics. *Proceedings of the 24th International Conference for the Psychology of Mathematics Education, 1*(1), 3–22. Hiroshima, Japan.

O'Leary, Z. (2004). *The essential guide to doing research.* London: Sage.

Ostermeijer, M., Boonen, A. J. H., & Jolles, J. (2014). The relation between children's constructive play activities, spatial ability, and mathematical word problem-solving performance: a mediation analysis in sixth-grade students. *Frontiers in Psychology, 5,* 782. doi: 10.3389/fpsyg.2014.00782

Paivio, A. (1971). *Imagery and Verbal Processes*. New York: Holt, Rinehart and Winston.

Piaget, J. (1951). *The Psychology of Intelligence*. London: Routledge.

Piaget, J., & Inhelder, B. (1958). *The Growth of Logical Thinking from Childhood to Adolescence*. New York: Routledge.

Piaget, J. (1970). Piaget's theory. In: P. H. Mussen (Ed.), *Carmichael's handbook of child psychology*. New York: Wiley.

Pimm, D. (1981). Metaphor and analogy in mathematics. *For the Learning of Mathematics, 1*(3), 47-50.

Pimm, D. (1987). *Speaking mathematically: Communication in mathematics classrooms*. London: Routledge.

Pinkernell, G. (2003). *Räumliches Vorstellungsvermögen im Geometrieunterricht: Eine didaktische Analyse mit Fallstudien* [Spatial ability in geometry lesson: a didactical analysis with case studies]. Hildesheim: Franzbecker.

Plath, M. (2014). Räumliches Vorstellungsvermögen im vierten Schuljahr. Eine Interviewstudie zu Lösungsstrategien und möglichen Einflussbedingungen auf den Strategieeinsatz [Spatial ability at grade four level. An interview-based study about solution strategies and possible influencing factors on strategy choice]. Hildesheim: Franzbecker.

Pólya, G. (1949/1980). On solving mathematical problems in high school. In: S. Krulik & R. Reys (Eds.), *Problem solving in school mathematics: 1980 yearbook* (pp. 1–2). Reston: National Council of Teachers of Mathematics.

Prediger, S. (2013). Darstellungen, Register und mentale Konstruktion von Bedeutungen und Beziehungen – Mathematikspezifische sprachliche Herausforderungen identifizieren und überwinden [Representations, registers and mental construction of meanings and relationships – identifying and mastering mathematics-specific linguistic challenges]. In: M. Becker-Mrotzek, K. Schramm, Thürmann, E., & H. J. Vollmer (Eds.), *Sprache im Fach – Sprachlichkeit und fachliches Lernen* (pp. 167–183). Münster: Waxmann.

Prediger, S., & Meyer, M. (2012). Sprachenvielfalt im Mathematikunterricht – Herausforderungen, Chancen und Förderansätze [language diversity in mathematics classroom – challenges, opportunities and support]. *Praxis der Mathematik in der Schule, 54*(45), 2–9.

Prediger, S., Renk, N., Büchter, A., Gürsoy, E. & Benholz, C. (2013). Family background or language disadvantages? Factors for underachievement in high stakes tests. In: A. Lindmeier & A. Heinze (Eds.), *Proceedings of the 37th Conference of the International Group for the Psychology of Mathematics Education, 4,* 49–56. Kiel, Germany.

Prediger, S., Wessel, L. (2011). Darstellen – Deuten – Darstellungen vernetzen: Ein fach- und sprachintegrierter Förderansatz für mehrsprachige Lernende im Mathematikunterricht [Represent – Interpret – Linking representations: a content and language integrated support for multilingual learners in mathematics classes]. In: S. Prediger, & E. Özdil (Eds.), *Mathematiklernen unter Bedingungen der Mehrsprachigkeit – Stand und Perspektiven der Forschung und Entwicklung* (pp. 163–184). Münster: Waxmann.

Pylyshyn, Z. W. (1973). What the mind's eye tells the mind's brain: A critique of mental imagery. *Psychological Bulletin, 80,* 1–25.

Radford, L., & Barwell, R. (2016). Language in mathematics education research. In: A. Gutiérrez, G. Leder, & P. Boero (Eds.), *The second handbook of research on the psychology of mathematics education. The journey continues* (pp. 275–313). Rotterdam: Sense.

Reeves Sanday, P. (1979). The Ethnographic Paradigm(s). *Administrative Science Quarterly, 24*(4), 527–538.

Richards, J. (1991). Mathematical discussions. In: E. von Glaserfeld (Ed.), *Radical constructivism in mathematics education* (pp. 13–51). Dortrecht: Kluwer.

Rost, D. H. (1977). *Raumvorstellung: Psychologische und pädagogische Aspekte* [Spatial ability: psychological and pedagogical aspects]. Basel: Beltz.

Rott, B. (2011). Problem Solving Processes of Fifth Graders: an Analysis. In: B. Ubuz (Ed.), *Proceedings of the 35th Conference of the International Group for the Psychology of Mathematics Education, 4,* 65–72. Ankara, Turkey.

Saussure, F. de (1916). *Cours de linguistique générale* [Course in General Linguistics]. Paris: Éditions Payot et Rivages.

Schoenfeld, A. H. (1983). *Problem solving in the mathematics curriculum: A report, recommendations, and an annotated bibliography.* Washington: Mathematical Association of America.

Scholnick, E. K., & Friedman, S. L. (1987). Reflections on reflections: What planning is and how it develops. In: S. L. Friedman, E. K. Scholnick, & R. R. Cocking (Eds.), *Blueprints for thinking: The role of planning in cognitive development* (pp. 515–534). New York: Cambridge University Press.

Schütte, M. (2009). *Sprache und Interaktion im Mathematikunterricht der Grundschule* [Language and Interaction in mathematics class at primary school]. Münster: Waxmann.

Schultz, K. (1991). The contribution of solution strategy to spatial performance. *Canadian Journal of Psychology, 45,* 474–491.

Schwank, I. (2003). Einführung in funktionales und prädikatives Denken [An introduction to functional and predicative thinking]. *Zentralblatt für Didaktik der Mathematik, 35*(3), 70–78.

Serbin, L. A., & Connor, J. M. (1979). Sex typing of children: play preference and patterns of cognitive performance. *Journal of Genetic Psychology, 134,* 315–316.

Sfard, A. (1991). On the Dual Nature of Mathematical Conceptions. Reflections on Processes and Objects as Different Sides of the Same Coin. *Educational Studies in Mathematics, 22*(1), 1–36.

Sfard, A. (1998). On Two Metaphors for Learning and the Dangers of Choosing Just One. *Educational Researcher, 27*(2), 4–13.

Siegler, R. S., & Jenkins, E. (1989). *How children discover new strategies.* Hillsdale: Erlbaum.

Smith, I. M. (1964). *Spatial ability, its educational and social significance.* San Diego: Knapp.

Spering, M., & Schmidt, T. (2009). *Allgemeine Psychologie kompakt* [General psychology compact]. Basel: Beltz.

238

Steinke, I. (2004). Quality Criteria in Qualitative Research. In: U. Flick, E. von Kardorff & I. Steinke (Eds.), *A Companion to Qualitative Research* (pp. 184–190). London: Sage Publications.

Strauss, A. (1991). *Creating Sociological Awareness.* New Brunswick: Transaction Books.

Strauss, A., & Corbin, J. (1998). *Basics of Qualitative Research: Techniques and Procedures for Developing Grounded Theory* (2nd Edition). Thousand Oaks: Sage.

Swain, M. (2005). The output hypothesis: theory and research. In: E. Heinkel (Ed.), *Handbook of research in second language teaching and learning* (pp. 471–483). Mahwah: Erlbaum.

Talmy, L. (1983). How language structures space. In: L. Herbert, J. Pick, & L. P. Acredolo (Eds.), *Spatial orientation: Theory, research and application.* New York: Plenum.

Taylor, H. A., & Tversky, B. (1996). Perspective in spatial descriptions. *Journal of Memory and Language, 35*, 371–391.

Teppo, A. R. (2015). Grounded theory methods. In: A. Bikner-Ahsbahs, C. Knipping & N. Presmeg (Eds.), *Approaches to Qualitative Research in Mathematics Education. Examples of Methodology and Methods* (pp. 3–21). Heidelberg: Springer.

Thomas, D. R. (2006). A General Inductive Approach for Analyzing Qualitative Evaluation Data. *American Journal of Education, 27*(2), 237–246.

Thurstone, L. L. (1938). *Primary mental abilities.* Chicago: The University of Chicago Press.

Thurstone, L. L. (1950). Some primary abilities in visual thinking. *Proceedings of the American Philosophical Society, 94*(6), 517–521.

van Hiele, P. M. (1986). *Structure and insight: A theory of mathematics education.* Orlando: Academic.

van Valin, R. D. (2001). *An Introduction to Syntax.* Cambridge: Cambridge University Press.

Vandenberg, S. G. & Kuse, A. R. (1978). Mental Rotations. A group test of three-dimensional spatial visualization. *Perceptual and Motor Skills, 47,* 599–604.

Vasta, R., Knott, J. A., & Gaze, C. E. (1996). Can spatial training erase the gender differences on the water-level task? *Psychology of Women Quarterly, 20,* 549–567.

Vygotsky, L. S. 1993. *Thought and language* (A. Kozulin, Ed. and Trans.). Cambridge: MIT Press. (Original work published in 1934)

Wagenschein, M. 1989. *Erinnerungen für Morgen – eine pädagogische Autobiographie* [Memories for tomorrow – a pedagogical autobiography]. Weinheim, Basel: Beltz.

Wessel, L. (2015). *Fach- und sprachintegrierte Förderung durch Darstellungsvernetzung und Scaffolding. Ein Entwicklungsforschungsprojekt zum Anteilbegriff* [Content and language integrated support by representation networking and scaffolding. A development research project about the concept of fractions]. Heidelberg: Springer Spektrum.

Winter, H. (1996). Mathematikunterricht und Allgemeinbildung [Mathematics lessons and general education]. *Mitteilungen der Gesellschaft für Didaktik der Mathematik, 61,* 37–46.

Witkin, H. A., Dyk, R. B., Faterson, H. F., Goodenough, D. R., & Karp, S. A. (1962). *Psychological Differentation.* New York: Wiley.

Wollring, B. (1998). Robert zeichnet und baut [Roberts draws and builds]. In: A. Peter-Koop (Ed.), *Das besondere Kind im Mathematikunterricht der Grundschule* (pp. 155–163). Offenburg: Mildenberger Verlag.

Wollring, B. (2012). Raumvorstellung entwickeln. Eine zentrale Forderung für mathematische Bildung [Developing spatial ability. A central demand for mathematics education]. *Fördermagazin, 2,* 8–12.

Zacks, R. T., & Hasher, L. (2002). Frequency processing: A twenty-five year perspective. In: P. Sedlmeier & T. Betsch (Eds.), *ETC. Frequency processing and cognition* (pp. 21–36). Oxford: Oxford University Press.

Zimmerman, B. J., & Pons, M. M. (1986). Development of a Structured Interview for Assessing Student Use of Self-Regulated Learning Strategies. *American Educational Research Journal, 23*(4), 614–628.

Printed in the United States
By Bookmasters